CRIME'S POWER

Anthropologists and
the Ethnography of Crime

Edited by
Philip C. Parnell
and Stephanie C. Kane

palgrave
macmillan

CRIME'S POWER

First published in 2003 by PALGRAVE MACMILLAN™
175 Fifth Avenue, New York, N.Y. 10010 and
Houndmills, Basingstoke, Hampshire, England RG21 6XS.
Companies and representatives throughout the world.
PALGRAVE MACMILLAN is the global academic imprint of the Palgrave
Macmillan division of St. Martin's Press, LLC and of Palgrave Macmillan
Ltd. Macmillan® is a registered trademark in the United States, United
Kingdom and other countries. Palgrave is a registered trademark in the
European Union and other countries.

1-4039-6179-4 hardback—1-4039-6180-8 paperback

Library of Congress Cataloging-in-Publication Data
Crime's power : anthropologists and the ethnography of crime / edited by
Philip C. Parnell and Stephanie C. Kane.
     p.   cm.
  Includes biibliographical references and index.
   ISBN 1-4039-6179-4—ISBN 1-4039-6180-8 (alk. paper)
   1. Criminal anthropology.   I. Parnell, Philip C.   II. Kane, Stephanie C.

HV6035.C74   2003
364.2—dc21   2003046020

A catalogue record for this book is available from the British Library.

Design by Letra Libre, Inc.

First Palgrave Macmillan edition: July 2003
10   9   8   7   6   5   4   3   2   1

Printed in the United States of America.

CONTENTS

LIST OF ILLUSTRATIONS

CRIME'S POWER

Philip C. Parnell

CRIME, LAW, AND THE STATE

ALTHOUGH CRIME AND OFFICIAL LAW ARE OFTEN directly linked in the views and parlances of citizens, researchers, the media, and the state, anthropologists conducting research on crime today show that social processes can affect crime and law in very different ways. Anthropologists, working as ethnographers and drawing on their knowledge of the locations of crime, present crime and criminalization as forms, experiences, and modes of expression used to manage the social nexuses of power. Such phenomena are especially dramatic in processes of social change when the nexuses of power are realigned and explored through revolution, democratization, migration, deindustrialization, the internationalization of capital and corporations, the rise and fall of dictatorships, the dissolution of states and empires, and war and its fading remembrance. The contributors to this volume show that crime is created as people brought together through these processes experience them differently, reasserting perceived essential differences among themselves (as gender, race, ethnicity, morality, and humanity) in changing circumstances. Both crime and law are created in culture as people negotiate social change. At the same time, crime and law may arise as separate culture-making processes and become linked to each other in a wide array of associations—for example, crime can be created even though the official law has not been broken or crime may be viewed as using rather than breaking state law. Anthropological research also raises the question of how crime's relationship to the state can be distinguished from its relationship to law. Though crime may be seen in the absence of lawbreaking, can the creation of crime today ever be separated from ways that states have exercised power in both local and international settings and continue to participate in the creation and distribution of power?

In this introduction, I consider themes in anthropological research on crime while focusing in particular on how societal processes are related to ways people use the criminal category in their everyday lives. As authors in this volume, as well as other anthropologists discussed here, focus on crime, they illustrate crime's ability to fracture and partition social nexuses and organize ways that groups interrelate. Crime does this in part through both symbolizing and obfuscating relationships, creating parsimony in processes of communication.

I initiate this essay with a brief general discussion that considers complex questions about crime by grounding them in two ethnographic examples that, in revealing interrelationships, work at odds with the power of the criminal category. Daniel Linger and Hilary Kahn, in this volume, detail two very different relationships between crime and the state. Linger shows state power traveling through crime directly into the everyday lives of people living on society's margins. Kahn looks more at structures of thought, following the construction of crime in myth as it was transformed under colonialism and works today to organize relations among diverse ethnic groups. Like all contributors to this volume, they illustrate crime's power in creating and controlling intersections between the translocal and the local.

Linger shows how the exercise of state power has been localized as violence against and by "criminals" in urban Brazil during the years that have followed the 1985 fall of Brazil's military dictatorship. Torture routines used widely by the state suffuse the larger society with what Linger calls a "suffocating authoritarian logic." Local police and the poor adopt the state's violent tactics, using them against the powerless who live on society's margins. As they are raped and beaten, powerless victims, such as women and men perceived to have violated moral gender roles, are turned into criminals, not through legal prosecution but as the violent punishments of the state are enacted on their bodies. In this way, as Linger explains, the woman who is raped in her neighborhood becomes a "criminal" in being the victim of an act of castigation. Originating in state exercise of political control, abusive and brutal ways of expressing power across social hierarchies turn powerlessness and vulnerability into a crime. Both the crime and the power that created it are violently inscribed on the victim's body as the practices of power originating in the state are translated into a sexual idiom. Linger observes, "the drastic way power is deployed is itself a horrific crime."

In the situation described by Linger, the state's past use of violence to control resistance in politically defined intersections of difference permeates a much wider range of differences across social hierarchies in general. Crime arises not through law but as and through ways people enact and recreate the state as a way of organizing differences that exist independently of the state apparatus. As Linger points out, crime created as the exercise of state power permeates society as that power is replicated locally to turn those perceived as different or weak into victims (criminals) by inscribing vulnerability on their bodies. Through linking ethnographic research on local acts of violence to historical knowledge of the Brazilian state, Linger shows the state's role in creating multiple criminal forms and expressions in the absence of legal regulation and within the most personal of relationships.

Yet the correspondences between the state and criminal actions are not always as direct as those observed by Linger. Hilary Kahn illustrates how Q'eqchi' who migrated from the Guatemalan countryside to the coastal town of Livingston come to understand relationships with "passersby" in the town by applying to them social categories the Q'eqchi' created during past "foreign intrusions," such as colonialism and the arrival of capitalism. The Garifuna, like the Q'eqchi', are officially classified as a Guatemalan indigenous group, and they preceded the Q'eqchi in Livingston. They are characterized by the Q'eqchi' through the mythical part-human, part-animal Q'eq, a creature symbolizing Otherness and criminality (he steals from the poor and gives to the rich). The Q'eqchi' created Q'eq out of Pre-Columbian myths while they were working as laborers on German-owned *fincas* (farms protected by the Q'eq for the Germans). The Garifuna, like Q'eq and unlike the Q'eqchi', are seen as black (*morenos*); but, most importantly, they violate a moral framework the Q'eqchi' developed as they worked on German fincas—one that prioritizes local community exchange with land and its owners. In contrast to the Q'eqchi', the Garifuna have migrated between Livingston and New York and Los Angeles; in the eyes of the Q'eqchi', some Garifuna are migrants to Livingston rather than its founders. Through migration, they have become participants in transnational economies and communities that are invisible to the Q'eqchi'. Because Garifuna moral exchanges cannot be seen in the absence of local community forms, the Q'eqchi' view them as "different, immoral, and natural transgressors" (although, as Kahn points out, Q'eqchi' and Garifuna moral frameworks are similar).[1] Today the Q'eqchi' have extended the same moral framework for understanding difference to other passersby—tourists, authorities of the state connected to foreign interests, guerillas, and Q'eqchi' youth and budding capitalists. Kahn explains that, for the Q'eqchi', "Criminals are defined by what they do, where they originate, who reaps the fruit of their labor, and with whom they collaborate or corroborate."

The Q'eqchi' criminal form, and its expression through myth, arises from political economies rather than legal interventions, though both its form and expression are clearly connected to relations of power. Foundations of today's criminalizations were created in state-based colonial transformations of local relationships of power through which Europeans became owners of land and controllers of economic resources while the Q'eqchi' became their laborers. Local myth that was transformed to represent these new relationships through a hybrid criminal creature remains a structure within which the movement of resources away from communities through today's different social nexuses is narrated as criminal behavior. Kahn explains: "Politics, economics, and religion continue to create mythic-historical structures that allow Q'eqchi' to innovate roles and understand crime." From a distance, equating the black Q'eq's criminality and moreno Garifuna migration might appear to be a product of racism; yet, in the ethnographer's closer examination of historical and contemporary local relations, notions of the criminal travel across time through economic changes rather than the conflation of skin color and criminality.[2]

Linger's and Kahn's analyses, as well as other anthropological studies of crime presented and discussed in this volume, suggest some clear distinctions between today's ethnographic research and that of earlier anthropologists who saw diversity in the ways societies were organized and "attempted to understand and then explain the ways in which different societies manage the serious wrongs that might endanger their peace and security" (Nader and Parnell 1982). Drawing on extensive ethnographic research, Bronislaw Malinowski, focusing on group-based consensus in relation to rules and writing in *Crime and Custom in Savage Society* (1926:99), defined crime as "the law broken" (which included violating taboos and challenging local hierarchies). Today, ethnographers, such as Laura Nader and Janine Wedel in this volume, noting the international spread of notions of law and crime that have arisen in Western rule of law societies, might ask to what extent can crime, when it is created out of non-legal situations, be locally viewed independently of Western legal frameworks and processes? Similarly, Linger illustrates that state use of punishment to distribute political power spreads throughout society where people use it to create criminal acts in various kinds of social hierarchies, leading one to ask whether acting *as if* the law has been broken has become a common way of enforcing and reinforcing various forms of powerlessness. A. R. Radcliffe-Brown (1933) located crime in how communities lacking law (as defined within Western legal systems) responded to their members' actions—in the collective expression that a moral code had been broken. Kahn's study, in which she notes the invisibility of moral relations among highly mobile capitalists, would lead one to rephrase Radcliffe-Brown's definition to state *as if* a moral code had been broken.

Recognizing the dynamic nature of culture and culture-making processes that take place at intersections of domains such as law, politics, economics, and kinship, and recognizing that relationships between cultures and places are ever subject to change, today's anthropologists are less inclined to seek essential relationships between law and crime, to search for universally shared notions of crime, or to seek explanations of criminal behavior through state-based definitions. Rather than creating taxonomies of allocated criminal categories and their treatments, anthropologists are focusing on the nature of crime at intersections of social and culturing processes—those processes that may change today's criminal categories or place them differently tomorrow. This approach not only distinguishes today's ethnographies of crime but, as Nader points out in this volume, it also distinguishes anthropological research on crime from much of the work in criminology.

Many anthropologists have assumed an epistemological approach to the relationship between the state and crime for several reasons. Whether one lives in a highly centralized or decentered state, the modern state, as it is experienced and imagined, is a readily available context through which people may perceive their problems and differences. As Linger points out, the state teaches people ways to deal with problems, opposition, differences, and resistance—at times, through ways it criminalizes and controls. And, the state may harness the power of the criminal toward its own purposes, as James Siegel (1998) illustrates through ethnography in Indonesia. Anthropologists have also traced some differences and intergroup conflicts in their ethnographic settings to actions of state-like structures in colonizing processes, from the spread of Western racial categories (American Anthropological Association 1998) to the British creation of criminal castes in India (Mayaram 1991) to creation of crime within mythical structures (Kahn in this volume). As a result of its growth, the modern nation-state, through sovereignty or interests, often participates in both local and international processes of change, influencing how change is perceived and controlled. And social and cultural anthropologists see such processes of change as locuses of crime's power: as examples, deindustrialization in the United States as it is linked to selling drugs and forming gangs (Bourgois 1995; Phillips 1999), the spread of international corporations into indigenous lands (Nader in this volume), the creation of "truth" in the resurgence of democracy in the Philippines (Parnell in this volume), or the movement of Western legal frameworks into former socialist states where boundaries to international aid and investment have weakened (Wedel in this volume).

Because the criminal category and criminalization are often ways of contextualizing and controlling difference, contexts of change that create new social nexuses—new intersections of the social relationships through which people perceive and symbolize differences and similarities—are contexts in which crime's power and its linkages to social and culturing processes can be revealed. They are places where ideological moorings are challenged and new opportunities for the reconstruction of differences are created. They are places where, as John and Jean Comaroff (1999:285) describe for a South Africa liberated from official apartheid, "realities appear more than usually fragile, fluid, fragmentary, and contested"—contexts in which boundaries of many sorts appear to be shifting and in which crime can play a central role in "repositioning fear and reassurance" (Siegel 1998:117). On the other hand, through processes of change, crime may be stripped of its power, as Vera Mark illustrates in this volume.

Crime, as Nader and Harry Todd (1978) detailed for law, may have varied roles and functions in society. Importantly, as contributions to this volume reveal, there is tension between the use of crime at social nexuses and the goal of the ethnographer. Crime and criminalization (as forms and expressions used to manage social forces) have the power to obfuscate and simplify relationships, processes, and experiences as well as render the roles and functions of the criminal category itself opaque. For example, crime may be used to locate or symbolize the tensions of change and conflict within and through a particular type of individual or group by drawing on the forces of religion, science, and the state; in the same way, crime may be used to justify existing distributions of power. Rather than treating crime as a fact of human pathology, anthropologists, no matter how horrific human behaviors have become, have looked beyond (and behind) the criminal category, finding its power in what it does not reveal or create—be it "the banality of evil," as the philosopher Hannah Arendt (1964) illustrates, or its displacement. What, then, do ethnographers see when they look into the locations of crime? What happens at social nexuses where crime arises? Through what processes do people situate criminal categories within their everyday lives? The following discussion looks more closely at such places and processes through contributions to this volume as well as the work of other anthropologists.

DISTINGUISHING THE POLITICAL AND THE CRIMINAL, THE FELLOW CITIZEN AND THE STRANGER

In Kahn's study, relationships between Europeans and a local population created through the international nexuses of colonialism and those that arise today as the same population confronts changes wrought by international migration and capital are brought together and mediated by frameworks of morality and crime. Linger's ethnography reveals that the state's criminal use of torture to create a political hierarchy travels widely into the margins of society as the Brazilian police and less powerful citizens reinforce hierarchies based on gender, politics, and views of morality by mimicking the state's uses of punishment. Crime interconnects state and local practices in the enforcement of perceived and desired hierarchies of power;[3] uses of violence that span extreme differences in power and class transect in the conflation of power and sex. Several contributors to this volume illustrate that criminalization becomes a way for those who represent state interests to distinguish among groups and things on the basis of types of relationships they can have with the state—it both creates and ranks group-based characteristics in relation to state goals. Examining a more conscious use of crime to structure society, JoAnn Martin illustrates in this volume how the Mexican state distributed criminality across social nexuses—relations between the state and population groups it distinguished as different—during the Mexican revolution to stabilize then shaky distinctions between political and criminal identities. Martin's analysis illustrates how criminal characteristics were used, with the help of science's positivism, to identify groups who, because they could not govern themselves, also could not enter into a proper relationship with the state. In the context of revolution, the foundation of distinctions between the criminal and the political is the notion of *governmentality*.

CRIME IN THE CONSTRUCTION
OF GOVERNMENTALITY

Moving from extensive ethnographic research in a town in Morelos, Mexico, into Mexican newspaper accounts written while the revolution was taking place (and before it was institutionalized), Martin explores the criminalization of a town (and now Mexican) hero—the Revolution's leader Emiliano Zapata—and his followers, the Zapatistas. In an interesting twist on locating objectivity, during the regime of Porfírio Díaz and the highly contested appropriations of peasant land he supported, criminalization of the revolution moves forward through the positivist intersections of "objective" journalism and science, rather than through state courts. Journalists present elite life as the proper intersection of self-governance and the state's management of society; as Martin observes, "Reporters' observations highlight elites as having cultivated the disciplinary routines of a free society." Criminality lies in the disruption of these routines—of "lives in progress." In newspaper accounts, the lives of peasants who, as criminals (revolutionaries), cross daily-life boundaries among classes, lead lives that lack detail and routine. They are collectively removed from lives in progress and, therefore, unable to experience disruption—they cannot be victimized by criminal acts. Reporters construct peasant villages suffused in silence (perhaps incomprehensible to those who have lived in Mexican peasant villages) to place them beyond the boundaries of governmentality. Martin points out that criminality and the tragic are established against the backdrop of "particular ways of life" rather than the "questionable" Mexican state. In contrast to the criminalization of peasant life, journalists, drawing on the character of the criminal as it was constructed at the time by criminal anthropologists and contrasting it with elite norms, criminalized Zapata by presenting him as insane—inherently criminal. Martin points out, however, that Zapata's self-presentations "emphasized an individual who could easily fit within the hegemonic elite order."

In Martin's account, the state and classes merge and diverge in battles over land, while criminality is used to barricade the doors of governance against an encroaching peasant class that is releasing prisoners and riding horses through sidewalk cafes. Martin points out that in Morelos today distinctions between the criminal and political remain unstable in accounts about Zapatistas.[4]

RELATING THROUGH THE CRIMINAL CATEGORY— POLITICAL RESISTANCE OR SUBJUGATION?

Anthropologists have often explored the political qualities of the criminal category through the framework of state domination and resistance to it. Don Kulick (1993), working in Papua New Guinea, finds that youth categorized as criminals (rascals) provide villagers with a discourse through which they can express dissatisfaction with and resistance to the disruptions caused by post-colonial, capitalist, and Christian influences (ibid.:10). Suzi Hutchings (1993) explains that some Aboriginal youth in Australia pattern their interactions with the larger society within criminalizing processes and define themselves in resistance to "dominant structures" (ibid.:356). They resist criminalization by using alcohol and drugs and through self-mutilation. Susan Phillips (1999) observes youth in the United States creating and interlinking gangs as new social forms that challenge social and cultural blueprints of the state. John Kelly (1990) examines how anomalies in British colonial thought about South Asians in Fiji, which he relates to British exploitation of Indian women, led to victims uniting around an alternative counter-hegemonic view of the world that facilitated their gaining what they were denied through centralized forces. Kelly argues that their resistance contributed to the eventual end of British colonialism in India. Anthropologists involved in AIDS research and intervention around the world have paid particular attention to the negative public health consequences created by the intersection of drug war politics and the criminal categories related to illegal drug use and prostitution (see review by Kane and Mason 2001).

David McMurray explores relationships between frontiers and criminality to present a different perspective on the political nature of state acts of domination in the relationships it structures by creating stages where crime can be enacted. He is standing along the North African border region that separates the Spanish enclave of Melilla from the Moroccon city of Nador, observing border guards repeatedly cuffing, slapping, kicking, and beating smugglers from Nador who beg for mercy. The guards, whom the smugglers view as criminals, confiscate some, but not all, of the smugglers' goods, then let them pass—these scenes are reenacted day after day as the smugglers connect the two cities and the markets into which they radiate. Given that smugglers are openly engaging in crime at the border, "the quintessential form of the state and its law," McMurray, in this volume, considers why the smugglers continually succeed at what they are doing. He does not find an adequate explanation in viewing the success of smugglers as resistance to a state that they indeed view as not having "a monopoly on reasons or a morally superior authority to determine for locals what is right and wrong." Rather, McMurray views this locus of crime as one place and process through which the state and its citizens interconnect to reaffirm shared concepts of the stylistics of power in the "weak" postcolonial state— a tattered symbolic state demanding rituals of subjugation from its cynical subjects through their need to make a living. The criminal aspects of border crossings are insignificant; citizens can smuggle as long as they recognize the state's right to enact its domination on their bodies—the ruler's right to rule and monopolize the use of violence, even in "stale," "hackneyed," "ordinary," and "vulgar" ways.[5] McMurray sees the border as a site for the reproduction of state power and, supporting the theoretical stance of political scientist Achille Mbembe (1992), observes, "popular recognition of the state's legitimate monopoly on ceremony and violence is what underlies most of the interactions between agents and subjects of the state." Criminalization and decriminalization combine at the legal border where poor males (the wealthier need not pass on foot) and soldiers repeatedly step onto the stage where violence and citizenship come together.

While suggesting that multiple forms of citizenship create multiple forms of crime, McMurray successfully conveys the sadness and indignity of the border where notions of crime are stripped of morality and suffused with the politics of power, an understanding of which unites those using violence to signify their roles as subjugators and subjects. Crime is just another context—here it is a place that symbolizes state rule—where the state and its citizens can assume their proper roles. It arises as the state intersects international economic networks that are mostly about economic exchanges for their participants, but are transformed by the state into opportunities for taking political tolls. The state's use of power to momentarily transform economic processes into political opportunities negates its own law, eliminates criminality through brutality, then turns what is—after the violence—ordinary trade into an opportunity for an insecure self-assurance of ceremonial power.[6] As McMurray points out, several anthropologists have noted borders and their crossings—the interstices of social, political, and economic systems—as places where crime's power may be used by differing parties to structure intersections in their own likenesses. And those who try to capture the power of such social nexuses risk criminalization, even as they use that process themselves. But the border between Melilla and Nador, stripped of the romance that can accompany resistance and the nobility that can accompany law, turn international competitions for power and territory into empty rituals of the state.

CRIME IN THE PARSING OF DISCOURSE AND THE STRUCTURING OF CONFLICT

In Martin's observations, journalists presented the state as needing citizens capable of self-governance (not all classes fitting this category) so that the government can focus on regulating relationships between people and material objects. In McMurray's example, citizens can violate state codes regulating their relationships with things, moving them across their legal boundaries, as long as they cooperate with state rituals of domination and subjugation at those boundaries. Anne Brydon and Pauline Greenhill illustrate in this volume that the criminal and his powers can be embodied in material objects. As these criminalized objects cross boundaries established for the criminal, they are positioned by hegemonic discourses linking crime to various interests and groups. The nature of crime, the nature of the objects, and the roles crime plays in the lives of community members—the search for a shared moral discourse about crime—are subjugated to constructing the boundaries of freedom around objects that, standing alone, enact no harm. Brydon and Greenhill focus on events in Winnipeg, a Canadian city, where the paintings of serial killer John Wayne Gacy are to be part of an art gallery show called

The Moral Imagination. But the "banal" depictions of clown-like figures are never shown; the art is categorized and confined as the brutal killer of young boys has been, disallowing confrontation both with the art (as the killer's representations), the nature of the killer's pathologies, and the complex roles that crime plays in the lives of urban residents.

Brydon and Greenhill, examining "how a criminal figure . . . mediates the production of social identities in a politicized debate over power and representation," illustrate the power of the criminal category to simplify not only the criminal—whose creative qualities are distinguished by citizens from those of artists—but also discourses about freedom and the politics of representation. Social networks brought together by the paintings—victims' rights groups, artists and their associations, the media, medical workers—are locked into the political discourse of controlling (rather than understanding) crime, as it is often fueled, in this case as well, by the tabloid media. As can happen through community policing and crime-watch programs (Darian-Smith 1993), the basis of the community's social nexuses is the threat those within it pose to each other—even when the threat is symbolized by an imprisoned killer and paintings that lack even the power of provocative art. Because the conflict over the paintings took place within hegemonic discourses about crime control—the harms of crime and the roles of censorship—the conflict was able to "reify" and "balkanize" relationships and failed to define a "community" of interest among victims' rights groups, artists, curators, and others. Even though the paintings, to past victims, represent the ways that crime has entered their lives, they cannot discuss the feelings such evocations engender—their sources must be hidden from view. Brydon and Greenhill reveal how an art gallery can be transformed by the groups crime brings together into both a religious ("temple") and legal ("courtroom") space. Considering the lost opportunity for developing a non-hegemonic discourse about crime (which Carol Greenhouse also considers in this volume) they observe: "The potential for suffering to be relived without being redeemed was kept alive. . . . The Gacy paintings perpetuated the marginalization of victims' concerns and demonstrated a curious blindness to the complex power of art."[7]

CRIME AS A CONFLATION OF THE LOCAL AND INTERNATIONAL

As Kahn notes, the ethnographer moving across numerous boundaries to learn about the dynamics and details of social processes and conflicts can be grouped among those whose ways of living violate local codes; from local perspectives, the ambiguous social relationships and forces that accompany the ethnographer may link with other "unknowns" that are categorized to symbolize what acts on daily life. In a self-reflexive return to her original field research in Turkey on village disputing, June Starr reveals in detail how identities and the out-of-place unite through the approaching unknown as war with Greece over Cyprus appears on the Turkish horizon. Starr, abandoned in the field by her husband and alone with her infant son, hires a young fisherman who speaks Turkish and English as her research assistant. An unaccompanied woman with a child and now in a relationship that also crosses local boundaries, Starr communicates and meets with several Turkish officials and becomes linked to other "foreigners" as she seeks to gain official permission to continue residency and research in Turkish villages. As Starr moves through these relationships, Turkish officials associate the "foreign" with threats on its borders and shape Starr's out-of-place identity through accusations of criminality—she is accused of smuggling antiquities, dynamite is found (placed) in her car by an official, and she is accused of spying (though she has permission from within the legal system to study courts).

As in Vera Mark's consideration of the life and identities of a French cobbler in this volume, Starr's insights from within the process (and, for her, experience) of criminalization reveals disjunctures between legal categories and what they symbolize. Is crime a means of limiting the circulation of knowledge—of fracturing its distribution across genders, nationalities, and class, for example—and positioning these identities in relation to each other through contexts created by the state? Or, was Starr, a woman moving and associating relatively freely (like Kahn and other passersby in Livingston who are not grounded in local community obligations), delinked from local notions of authority, her criminalizations locating her within governmental control rather than expressing notions of the law violated—a dynamic Carol Greenhouse considers in this volume? On the other hand, were the accusation-generated meetings with authorities exaggerated rituals of domination and subjugation, as McMurray's ethnography suggests? Did Starr's criminality lie in violating the codes of these processes? Was Starr's eventual exit from Turkey related more to the conflict over Cyprus and the unknowns of her (foreigner) status than more specific relationships and identities? Numerous nexuses of crime unfold and interlink as Starr the ethnographer moves through her days and Turkey as a Western woman out of place. The challenges of doing ethnography of conflict appear as professional and personal hurdles that Starr overcame as they intermingled with local and national forces of control.

CRIME IN THE RECOMPOSITION OF KNOWLEDGES

As observed above in examples ranging from Livingston to Winnipeg to Turkey, criminalization creates barriers to communication. Attempts to avoid the power of the criminal category may also erect such barriers. Laura Nader, in this volume, looks across domestic, international, and academic landscapes to observe the shifting placement of the criminal category. She finds it can be misplaced and displaced through disrupting information flows that would reveal relationships among perpetrators, their harms, and their victims in situations of corporate homicide, toxic waste pollution, and corporate seizure of indigenous lands. Disjunctures in knowledge can weaken the abilities of victims to both have and protect their rights: judges can separate interrelated cases, making it difficult for juries to see how the harmful actions of large corporate bureaucracies are interrelated; scientific knowledge and public knowledge about the harms chemicals do may be out of synch with the suffering they cause—corporations can strongly influence the dissemination and development of that knowledge; common criminological conceptions of space may hinder the abilities of victims to seek redress against the harms of multinational corporations; and the power of corporations to influence the legal and popular allocation of rights among humans and the material (such as chemicals) may dislocate populations, change the local cultural meanings of bodies contaminated by industrial toxins, and criminalize those who, in expressing their knowledge, resist privatization of indigenous lands. Noting links between local practices in developing knowledge, the ability to establish "interrelatedness" between actions and their effects, and reactions to harms to communities, Nader draws on her ethnographic research in Mexico to observe: " . . . in societies where communities are in closer touch with issues of survival, the notion that people who endanger the public health of communities should be held responsible is much more widespread . . . [for the Zapotec of Oaxaca] endangering the interests of the Commons is among the most serious cases that are referred to the court."

Addressing a contradiction in Western legal thought and action and, here, a barrier to rather than a result of placing the criminal category, Nader states, "There is no rationality in the absence of interrelatedness." In the case of corporate harms, disjunctures in knowledge are created by the powerful to limit the linking of criminality and the powers of the criminal category to their acts. As corporations guard the line between criminal and civil law, the absence of the criminal category, more frequently applied to the poor for much lesser harms than those sometimes perpetrated on communities by corporations, becomes an expression of societal power: "Clearly, use of the criminal category, application of criminal sanctions, and, apparently, the public's general perception of the seriousness of an offense, all are mediated by the diacritics of power in the relationship between the perpetrator and the victim." Nader observes that the expanding interventions of today's multinational organizations increasingly interrelate people who differ in their knowledge of how multinationals work as well as in ways of organizing knowledge, in power, and the ways they distinguish between the civil and criminal. Such differences will increasingly hinder abilities of victims to locate responsibility in accessible legal forums and find redress for harms.

CRIME AT THE INTERSECTIONS OF POLITICAL CHANGE: THE UNKNOWN STATE

Looking across societies in the context of corporate power, Nader observes, "in the arena of international interactions, significant cultural barriers exist in the field of legal ideas." Through such barriers, the past cold war East-West opposition of communism versus capitalism could be transformed today into a myopic relationship of cops and robbers; as Janine Wedel writes in this volume, "When Western institutions export the rule of law, they often impose assumptions on Eastern European societies that may not coincide with indigenous views of how state and private should interrelate, which are nuanced."[8] Drawing on extensive ethnographic research in eastern Europe and her studies of Russian state-private organizations, Wedel reveals several social nexuses becoming categorized as criminal or corrupt as Westerners use crime to assert authority over ways eastern Europeans govern and do business, and as eastern Europeans use Western notions of the *mafia* to symbolize sources of uncertainties and the unexpected in postsocialist states. These two processes combine in criminalizing what Wedel calls "flex organizations": "organizational structures set up by informal groups to cross-cut and mediate institutional spheres . . . [that have an] . . . adaptable, chameleon-like, multipurpose character." Flex organizations traverse and blur boundaries between state and private, bureaucracy and market, and legal and illegal that are well-bounded in the rhetoric of Western public administration. Complicating Western views of such intersections of social networks as illegal is the fact that, in eastern Europe, violating the law is not necessarily equated with criminality: "The law is often used for ad hoc purposes, bargaining, and extracting advantages." Importantly, these social membranes and nexuses are not always visible, which is evident as Wedel, citing legal analyst Jan

Steanowicz, points out that today "Some 30 percent of the Polish budget lies somewhere between the private and the state sector."[9]

Gaining knowledge of how to intersect political, economic, legal, private, and governmental spheres was, under communism, a way of surviving within the uncertainties of a state-controlled highly bureaucratized economy; in these circumstances, a state and society dichotomy developed as citizens struggled to assert authority in their lives. Unofficial networks linking society to state bureaucracies played an important role in the reform processes of the 1990s, and today serve as "European counters to centralization, of political and economic control in the hands of a few local and international elite players." Wedel details the nexuses of private and public organizations in Russian and eastern Europe, arguing that Western vocabularies of corruption and crime used to promote rule of law and that permeate the social sciences, policy making, and popular discourse lead to misunderstanding eastern European social systems and their criminalization. Based on culturally grounded private-public distinctions, Western concepts of corruption hinder development of a non-legalized discourse for understanding and relating to public-private institutional networks that have become normalized in Russia and eastern Europe. Eastern European and Russian use of the term mafia exacerbates this myopia. Wedel, noting similarities with John and Jean Comaroff's (1999) study of people responding to change today in South Africa through witchcraft accusations, illustrates the misleading use of the borrowed concept of mafia by eastern Europeans to control sources of uncertainty and explain growing divisions in wealth.

CRIME IN DIMENSIONS OF DEMOCRATIC PROCESSES

Criminalizing processes enter relations among intergroup networks of the urban poor, without the invocation of law, as the poor link to governmental, religious, and legal institutions to form localized versions of the state in Phil Parnell's ethnographic study in this volume, which emanates from a sprawling squatter settlement in Metropolitan Manila. But there, the criminal category is used among neighbors and former allies rather than across international divides and distances between those who are making out and those who are losing out. As in eastern Europe and South Africa, as well as in Turkey facing the possibility of war and revolutionary Mexico, crime becomes momentarily salient in Manila intergroup relations while change engulfs the Philippines following the end of dictatorship through the People Power Revolution of 1986, and the resurgence of democracy. As the discourses of democracy permeate governmental relations, urban poor people's organizations that have supported democracy and practice it as they seek land reform are empowered through the force of homology—through governmental restatement and universalization of their organizational practices as national ideology (as with journalistic depictions of elites in Mexico). This process is accompanied by another one—iconization—through which more bureaucratically organized networks—land syndicates—that invoke the law to redistribute urban land along lines drawn under Spanish colonialism are simplified, their similarities to democratic networks rendered opaque, and their associations with past and present "misuses" of the law forefronted. Through these processes the criminal category becomes salient and gives direction to social intersections where it previously was little used; syndicates become viewed as criminal by some of their urban neighbors and opponents as well as within governmental agencies, although no legal charges are filed against them. One became the target of official state military force.

Syndicates are criminalized in part because they associate with law— rather than because they break it—to impose solutions on conflicts over resources among competing intergroup networks, as happened during past impositions of colonialism and under the martial law of the past dictatorship. On the level of the state, the criminal category is applied as the power of intergroup networks to unite numerous sectors of Philippine urban society with state organizations increases. It is placed on socially isolated components of syndicates and on larger syndicate organizations as a means to control their networking potential. Urban poor networks that compete with syndicates for control over urban land use organizational strategies similar to those of syndicates—all of the intergroup networks, as Wedel describes for Russia, blur public and private spheres. Because survival is based on creating interrelatedness through these networks, the criminal category assumes only a shaky legitimacy (as Martin observes for separations of the criminal and political in Mexico). Crime becomes a component of social relations as long-standing local processes of surviving through associating (in various ways) intersect with political and ideological changes in the Philippine state and nation. Among the urban poor of Manila, as among migrants in the town of Livingston, interest groups in Winnipeg, and eastern European citizens of postsocialist states, crime as a category intervenes in communication across groups that share common goals and practices, redirecting it into the language of conflict.

CRIME AS POWER CONFINED

The importance of such multiple-purpose intergroup networks linking the local to the state to survival in contexts of change is illustrated in the United

States by their absence, which anthropologists have related to criminalization of local populations isolated by processes such as deindustrialization. Philippe Bourgois (1995) describes drug users and sellers of Puerto Rican descent in New York City as bereft of cultural capital in the growing service economy—they can no longer use ways of networking received from Puerto Rico or masculine ways of relating on the job that might have been appropriate among laborers. Just as Susan Phillips (1999) points out for gangs in Los Angeles, power becomes isolated to the local as people are cut off from access to cultural capital that would help them gain employment. Their marginalization can then contribute to their criminalization. Mercer Sullivan (1989) also describes relationships between changing social nexuses of deindustrialization and criminalization as the U.S. labor market becomes segmented, serving to "divide jobs and workers on the basis of race, ethnicity, citizenship, age, and sex" (ibid.:225). The dislocations of such changes are exacerbated as notions of citizenship become linked to consumerism. In Parnell's study, intergroup networks cutting across economic and political differences seem to generally mitigate against the formation of localized groups and networks that are viewed as criminal. In Wedel's analysis, they are clearly accepted forms of doing business and politics, providing useful linkages between past mechanisms (networks) of survival and changing political and economic conditions.

CRIME AS ITS POWER WANES

The salience of the criminal identity in Manila, observed by Parnell in the context of the forces and relationships that made it somewhat credible, faded with time and the institutionalization of change; as Vera Mark illustrates in this volume, just as changing circumstances can undo the intersections that locate crime within individuals and groups, relationships transformed by crime can linger into the future. Crime, once created, can move across relationships through its new political allocations; and criminal identities can invite reconfiguration as the disappearance of circumstances reveals the ambiguities of facts. In writing about the life of a French cobbler, small shopkeeper, and poet who was found guilty of World War II collaborationism ("the ideological acceptance of fascism") in a civil court (he was later granted amnesty), Mark reveals the vagaries of conducting an ethnography of crime, even though she knew the cobbler, his family, and the writings through which he forever sought to establish his innocence: perhaps his conviction was a punishment for his daughter's traversing social class boundaries through marriage; or perhaps he paid the price of being powerless and staying in his community, unlike others suspected of collaborating who fled or could afford powerful lawyers.

Mark, drawing on the work of others who have studied collaborators, notes that "retribution for collaboration differed according to groups, such that experts, businessmen, and bureaucrats survived almost intact . . . while the purge struck much harder at men of words," such as journalists and intellectuals. Importantly, Mark realizes that private life is a barrier that not even the ethnographer can cross. State regulations and French law restrict access to World War II archives and protect the identities of those whose lives are revealed within them; amnesties and pardons legally erase the political past. Searching circumstances, Mark finds support for the accusations of collaborationism in the cobbler's education, past military service, political affiliations, and the economic anti-semitism of southwestern France at the time—but these characteristics do not distinguish him from many other men. Through other stories, Mark also realizes that being exceptional in various ways, such as violating community notions of solidarity, could render one the target of criminalizing processes that had great currency at the time—even a resistor was shunned for surviving while his comrades died. The pain of criminalization blocks communication about events within the family; the issue of the French state's complicity, versus the individual's, shifts the focus of memory and access to written records. And, in French communities, the secret, "an integral part of the local linguistic economy," situates "individuals across time." Mark concludes, "Gaps between the written and oral, the past and present, and words and actions make it difficult to determine [the cobbler's] guilt."[10]

Lessons about crime may be hard to learn if they are to be based on a hierarchy of facts. Is the criminal category supposed to stand alone as the lesson itself? The ethnographer, living within the circumstances, crossing the social intersections, and traveling through the nexuses of power that become identified, circumscribed, and invigorated by criminal categories—and that in turn give power to the criminal category—can, at times, see beyond what criminalization obscures to know the more complex and complicated individuals and human contexts that collectively live more broadly than legal roles and state inscriptions. Some behaviors labeled criminal that are committed by individuals, groups, and the government are horrific, as are some not so labeled; but, given the nature of the criminal category and its broad use and production within and outside officially legal forums, can it contain or prevent such behaviors? As the ethnographer looks within and beyond crime and criminalization, how might an ethnography of crime—a process of revelation—differ from ways that criminalization and the processes that accompany it simplify identities and relationships, interrupt discourse within conflict, reinforce hierarchical disjunctures, and obscure the nature of differences?

CRIME WITHOUT AUTHORITY

Carol Greenhouse, moved by her experience of "unfreedom" while visiting a New York jail, addresses the above question of an ethnography of crime while "re-reading Durkheim," in which she contemplates his meaning of the "collective" or "common conscience" and the ethical nature of objectivity. Writing in this volume, she finds that the role of the ethnographer can differ, indeed should differ, from the role of the state and law in response to crime. Importantly, in relation to organic solidarity, Durkehim's "term for associations that span normative communities" (rather than unanimity), Greenhouse argues Durkheim's central thesis is "that there is a direct relationship between organic solidarity and individual autonomy, given the way the recognition of multiple normative systems heightens an individual's self-awareness as a moral actor and buffers or mediates the internalization of any one normative discourse arising from the pronouncements of judges or other public authorities." The dynamics of the criminal category differ from those of organic solidarity, in which "the public recognition of social and normative diversity fosters an individual's awareness of him- or herself amongst others, and accordingly, to effective moral reflection and deliberative change." Categories of crime, on the other hand, in Durkheim's reading, "do not derive from some universal scale of evils or public consensus as to interests, but from the way systems of authority make themselves known and maintain themselves 'in some way transcendent'"—"The interests of authority and its needs for self-legitimation determine *crime*, then, not the nature of the acts in question." In this way, law falls outside the collective or common conscience and, for Durkheim, "it is social science—not law, politics, or philosophy—that articulates the collective conscience as a register for public discourse." But how can social scientists reveal the collective conscience without articulating the interests of authority? Greenhouse explains that:

> Durkheim's articulation of the sociological method takes the form of direct appeals to the reader's capacity for fairness and reflection. It would seem that if social science is objective it is not because objectivity is some inherent property of the sociologist's activity, but because it defines a relation between the sociologist's discernment and alternative modes of knowledge. Scientific objectivity, in other words, is contingent on ethical

dialogues and debate, on ethical judgments rooted in social convention and personal experience yet open to revision in the light of knowledge of the vagaries of the human condition.

To place Greenhouse's more elaborated exploration of an ethnography of crime in relation to the preceding discussion in this introduction (but, alas, without her eloquence), such an ethnography differs from state and legal contexts for understanding crime, and subsequent roles of the criminal category, through revealing rather than obscuring the various viewpoints that constitute communities and societies in ways that others may see them and consider them fairly. In this way, the collective conscience, in which the individual connects with the numerous normative spheres of society and develops the moral self, is enlarged. As Greenhouse explains, the process of becoming the moral self is a process of gaining freedom. And part of this freedom will lie in expanded communication across differences. In other words, articulating without judging what the criminal category obscures is an ever expanding process of developing moral selves who, "stripped of the syntax of social judgment," can better recognize ways that criminal categories—"originating not in the inherent criminality of particular acts or in the substantive public interests, but in the categorical opposition . . . between approval and reproof"—limit discourse through authoritative articulations of good and bad. An ethnography of crime, in communicating the knowledge that crime as a category masks, can also seek to create, for ethnographers and their readers, the sense that one has considered something fairly not for the purposes of gaining authority but to invigorate and enlarge the universe in which individuals articulate with its differences to locate the self. Greenhouse states, "For each of us, the need for social science may be different, but from that instant of grasping the objective element, we can be drawn through it into a common conversation."

THE INTERSECTION OF CRIME AND ETHNOGRAPHY

The preceding discussion suggests that crime can be used to distribute power in a range of intergroup social processes with or without the invocation of law. Crime as a culture-making process is used to shade perceptions in and of the social experience; in doing so, it provides experiences where societies' groups intersect with organization and structure. In the context of revolution, crime is a decidedly political category that equates ways of life with the right to govern through the state. On the other hand, in the process of democratization, crime only indirectly structures intergroup relations as part of the process of universalizing social authority. Sometimes, even those symbols that evoke thinking about crime can structure and confine thought in processes of communication and conflict. In carrying this power, crime becomes fungible with other categories that are products of conflicts in which differences have been essentialized. And crime can transport such disjunctures throughout processes of change as people use it to describe and control what appears to be out of place and unexpected. Transportable across social nexuses and interchangeable with other categories expressing difference, the criminal category can facilitate the movement, spreading, or oscillation of differences across political, legal, and criminal domains.

And, as the social nexuses at which crime is used and on which it is placed dissolve in the vicissitudes of time, crime can survive as a memory loosened from what once seemed more certain while it remains a catalyst for the partitioning of knowledge. Indeed, since constructing the criminal may depend on being able to see ways that people, behavior, and consequences are interrelated, some can escape the criminal category through the power to control knowledge of how people make things happen. In allocating power across dimensions of the human experience, the criminal category, at the same time, partitions knowledge of that experience. Reconstituting the communication that criminal categories can fracture can then only further the goal of understanding what it means to be human, fair, and moral in a world where differences are shared.

The criminal category is, then, a good place to hide things. It can effectively signify and create disconnections in communication across the groups, networks, states, and other relationships through which people symbolize themselves and others. It is part and parcel of the culturing and structuring processes of the experience of change, highlighting only portions of the new while placing them in the frameworks of the past. But as it reveals or delegitimizes some forms of interrelatedness and ways of communicating, it can also inspire new political discourses. Creating disconnections out of interconnections, moving across social nexuses as it is placed variously by the powerful and the powerless—the placing of the criminal category is dependent on the ways we see power working in relationships. Anthropologists often locate the origins of such cultural knowledge in the articulations of intergroup relationships and how humans and their bodies are used to symbolize them. Crime is often a part of that process. And, often, the criminal category seems to work at cross-purposes with ethnography, rendering invisible the meanings, processes, and interconnections in other peoples' lives that the ethnographer works to see. In this sense, crime may be a locus of ethnographic revelation, signifying the people, places, and processes through which ways of knowing and not knowing are interconnected. As the ethnographer seeks familiarity and communication, the criminal category reveals ways that people work against them as they create, receive, and try to move the boundaries of knowledge. The criminal category is then a barrier worth pushing against through ethnographic practice—what lies behind it may well be worth looking for.

ACKNOWLEDGMENT

I would like to thank Stephanie Kane for her comments on drafts of this introduction.

NOTES

1. A similar dynamic is analyzed by Sally Merry (1981) conducting ethnography in Dover Square, a low-income housing development in the United States. Merry proposes that "crime serves as an idiom for expressing and legitimating the fear of the strange and the unknown" (ibid.:14). She explains that residents of Dover Square fear strangers, the disorderly, the morally repugnant, and that which is culturally unfamiliar; social distance that accompanies ethnic differences may create a sense of danger that is then expressed as fear of crime. When this dynamic develops along lines of class and power, fears that accompany those differences may facilitate a dominant group's use of the criminal category: "I argue that the fear of crime serves as a way of rationalizing and legitimizing increased control of the subordinate populations who are labeled as dangerous" (ibid.: 220) (see also Sinha-Kerkhoff 1998).

2. Rachel Tolen's (1991) research on colonial British criminalization of Indian groups illustrates that the state may also integrate the criminal category into its technologies of control as it regulates boundaries of difference.

3. Use of violence to enforce gender boundaries and denigrate the victim's status is also illustrated in studies of domestic violence (see, for example, Connolly 2000; McGillivray and Comasky 1999; and Raphael 2000).

4. Ethnographic research by several anthropologists (Borneman 1998; Siegel 1998; Kelly 1990; Scully 1989) indicates the criminal category can be an expression of who controls the allocation of power across symbols, hiding what is really going on as groups negotiate, construct, and deconstruct inequalities in access to commonly valued material resources and forms of honor. In these contests, who is the criminal and who is the hero may shift across identities over time

5. The anthropologist John Devine (1996), conducting ethnography in inner city schools of New York, notes the use of surveillance in schools as a means of using the bodies of students to subjugate. There, this subjugation contributes to student use of violence (see also Tolen 1991; Cohn 1989; Pinney 1989; Guha 1983).

6. McMurray also notes in his study that the border ceremony reaffirms smuggler acceptance of lower social status, which could function to reassure their availability as part of a low-wage population. Tolen (1991) also notes state use of the criminal category to organize human sources of production.

7. John Kelly (1990) studying crime in Fiji under British colonialism and Don Kulick (1993) studying young criminals in Papua New Guinea note the role of crime in the creation of discourse, but, in these cases, crime and use of the criminal category help to engender an alternative to hegemonic discourse.

8. The dynamics noted by Nader through the growing overlap of corporations and governance and the spread of international corporations, and by Wedel through the spreading use of Western legal frameworks to replace former political oppositions in the international arenas of politics and economies, has been noted in the past, in relation to crime, by anthropologists studying plural systems of law within states. Several of these studies have been related to use of the criminal category of *thief* where there are differing state and local views on the movement of material goods. Studies of livestock displacement by Julio Ruffini (1978), Charles Cutshall (1982), and Michael Herzfeld (1988) illustrate the difficulty that can arise in drawing analytic distinctions between crime and law in situations of contact between distinctively different socioeconomic systems marked by disagreement over when to apply the category of crime. Additionally, Michael Herzfeld's (1988) study of Cretan theft illustrates that even within groups great variability and individual agency can arise in the application of criminal labels.

9. Susan Phillips (1999) draws on ethnography to place gangs in California in similar organizational frameworks: In her analysis they are located in the shadows between success and failure, in systematic persecution and exclusion from the larger society's resources and other expressions of power—"betwixt and between" cultures in "the fragmented, a historic, placeless urban worlds of late capitalist consumer culture" (ibid.: 73). She likens gangs to what the anthropologist Victor Turner (1969) called antistructure—the antithesis of the larger system developing in a self-reinforcing relationship with that system. She also relates their structures to social systems based on segmentary oppositions described by anthropologists working in non-state situations (see, for example, Evans-Pritchard 1940; Fortes and Evans-Pritchard 1940; and Waterbury 1970).

10. Mark also notes that barriers to understanding what happened during World War II can hinder understanding of responsibility within the processes and relationships of genocide. For further discussion of these processes and debate over state versus individual responsibility see, for example, Browning 1992a, 1992b; Goldhagen 1996; Kuper 1981; Svaldi 1989; Besteman 1999; and Gordon 1992.

BIBLIOGRAPHY

Abraham, Margaret
2000   Speaking the Unspeakable—Marital Violence among South Asian Immi-
       grants in the United States. New Brunswick, NJ: Rutgers University Press.
Altman, Irwin and Joseph Ginat
1996   Polygamous Families in Contemporary Society. New York: Cambridge Uni-
       versity Press.
American Anthropological Association
1998   AAA Statement on "Race." Anthropology Newsletter. September: 3.
Appadurai, Arjun
1990   Disjuncture and Difference in the Global Cultural Economy. Public Culture
       2:1–24.
Arendt, Hannah
1964   Eichmann in Jerusalem: A Report on the Banality of Evil. New York: Penguin
       Books.
Arnold, David
1979   Dacoity and Rural Crime in Madras, 1860–1940. Journal of Peasant Studies
       6(2): 140–167.
1984   Criminal Tribes and Martial Races: Crime and Social Control in Colonial
       India. Paper presented at a seminar entitled "Comparative Commonwealth
       Social History: Crime, Deviance, and Social Control," University of London
       Institute of Commonwealth Studies, London, December 4.
Asad, Talal
1975   Two European Images of Non-European Rule. *In* Anthropology and the
       Colonial Encounter. Talal Asad, ed. Pp. 103–118. London: Ithaca Press.
Bachman, Ronet
1991   The Social Causes of American Indian Homicide As Revealed by the Life Ex-
       periences of Thirty Offenders. American Indian Quarterly 15(4): 469–492.
Banfield, Edward
1970   The Unheavenly City. Boston: Little, Brown.
Ben-Yehuda, Nachman
1992   Criminalization and Deviantization as Properties of the Social Order. The
       Sociological Review 40(1): 73–108.
Besteman, Catherine
1999   Unraveling Somalia: Race, Violence, and the Legacy of Slavery. Philadelphia:
       University of Pennsylvania Press.
Blok, Anton
1974   The Mafia of a Sicilian Village 1860–1960: A Study of Violent Peasant Entre-
       preneurs. Oxford: Basil Blackwell.
Bohannon, Paul
1957   Justice and Judgement among the Tiv. London: Oxford University Press.
1960   African Homicide and Suicide. Princeton, NJ: Princeton University Press.
1965   The Differing Realms of the Law. American Anthropologist 67 (6, part 2): 33–42.

Borneman, John
1998   Retribution and Judgment: Violence, Democratic Accountability, and the In-
       vocation of the Rule of Law. Ethnologia Europaea 28(2): 131–149.
Bourdieu, Pierre
1977   Outline of a Theory of Practice. Cambridge, UK: Cambridge University
       Press.
Bourgois, Philippe
1995   In Search of Respect: Selling Crack in El Barrio. Cambridge, UK: Cambridge
       University Press.
Browning, Christopher
1992a The Path to Genocide: Essays on Launching the Final Solution. Cambridge:
       Cambridge University Press.
1992b Ordinary Men: Reserve Police Battalion 101 and the Final Solution in
       Poland. New York: HarperCollins.
Caldeira, Teresa
2000   Crime, Segregation, and Citizenship in São Paulo. Berkeley: University of
       California Press.
Chin, Ko-Lin
1990   Chinese Subculture and Criminality: Non-Traditional Crime Groups in
       America. Westport, Conn.: Greenwood Press.
Cicourel, Aaron
1967   The Social Organization of Juvenile Justice. New York: Wiley.
Cohen, Albert K.
1955   Delinquent Boys. New York: Free Press.
Cohn, Bernard S.
1989   Cloth, Clothes, and Colonialism: India in the Nineteenth Century. *In* Cloth
       and the Human Experience. A. B. Weiner and J. Schneider, eds. Pp.
       303–354. Washington, D.C.: Smithsonian Institution Press.
Collier, Jane
1973   Law and Social Change in Zinacantan. Stanford, CA: Stanford University Press.
Comaroff, Jean and John L. Comaroff
1999   Occult Ceremonies and the Violence of Abstraction: Notes from the South
       African Postcolony. American Ethnologist 26(2): 279–303.
Conley, John and William M. O'Barr
1998   Just Words: Law, Language, and Power. Chicago: University of Chicago
       Press.
Connolly, Deborah R.
2000   Homeless Mothers: Face to Face with Women and Poverty. Minneapolis:
       University of Minnesota Press.
Crummey, Donald, ed.
1986   Banditry, Rebellion, and Social Protest in Africa. London: James Currey.
Cutshall, Charles R.
1982   Culprits, Culpability, and Crime: Stocktheft and Other Cattle Maneuvers
       among the Ila of Zambia. African Studies Review 25(1): 1–26.

Daly, Martin and Margo Wilson
1988 Homicide. New York: A. de Gruyter.
Darian-Smith, Eve
1993 Neighborhood Watch: Who Watches Whom? Reinterpreting the Concept of Neighborhood. Human Organization 52(1): 83–88.
Davis, Mike
1990 City of Quartz: Excavating the Future in Los Angeles. London: Verso.
Day, Sophie, Evthymios Papataxiarchis, and Michael Stewart
1999 Lilies of the Field: Marginal People Who Live for the Moment. Boulder, CO: Westview Press.
Derrida, Jacques
1976 Of Grammatology. Gayatari Chakravorty Spivak, trans. Baltimore: Johns Hopkins University Press.
1997 Deconstruction in a Nutshell: A Conversation with Jacques Derrida. John D. Caputo, ed. and commentator. New York: Fordham University Press.
Devine, John
1996 Maximum Security: The Culture of Violence in Inner-City Schools. Chicago: University of Chicago Press.
Dickeman, Mildred
1979 Female Infanticide, Reproductive Strategies, and Social Stratification: A Preliminary Model. In Evolutionary Biology and Human Social Behavior. Napoleon Chagnon and William Irons, eds. Pp. 321–367. North Scituate, MA: Duxbury Press.
Durkheim, Emile
1933 Emile Durkheim on the Division of Labor in Society. George Simpson, trans. and commentator. New York: Macmillan.
Evans-Pritchard, E. E.
1937 Witchcraft, Oracles and Magic among the Azande of the Anglo-Egyptian Sudan. Oxford: Clarendon Press.
1940 The Political System of the Anuak of the Anglo-Egyptian Sudan. London: Lund Humphries & Co., Ltd.
Fallers, Lloyd A.
1969 Law without Precedent: Legal Ideas in Action in the Courts of Colonial Busoga. Chicago: University of Chicago Press.
Fleisher, Mark S.
1995 Beggars and Thieves—Lives of Urban Street Criminals. Madison: The University of Wisconsin Press.
Fortes, Meyer and E. E. Evans-Pritchard, eds.
1940 African Political Systems. London: Oxford University Press.
Foucault, Michel
1979 Discipline and Punish: The Birth of the Prison. New York: Vintage Books
Fuller, Lon
1969 The Morality of Law. New Haven: Yale University Press.

Geertz, Clifford
1999 "The Pinch of Destiny": Religion as Experience, Meaning, Identity, Power. Raritan: A Quarterly Review 18(3): 4.

Girard, René
1977 Violence and the Sacred. Patrick Gregory, trans. Baltimore: Johns Hopkins University Press.

Goldhagen, Daniel Jonah
1996 Hitler's Willing Executioners: Ordinary Germans and the Holocaust. New York: Alfred A. Knopf.

Gluckman, Max
1959 The Magic of Despair. The Listener, April 29, 1954: 724–724, 737.
1963 Order and Rebellion in Tribal Africa. London: Cohen and West.

Gordon, Robert J.
1992 The Bushman Myth: The Making of a Namibian Underclass. Boulder: Westview Press.

Gramsci, Antonio
1957 The Modern Prince and Other Writings. New York: International Publishers.

Greenhouse, Carol
1986 Praying for Justice: Faith, Order, and Community in an American Town. Ithaca, NY: Cornell University Press.

Guha, Ranajit
1983 Elementary Aspects of Peasant Insurgency in Colonial India. Delhi: Oxford University Press.

Gulliver, P. H.
1971 Neighbors and Networks: The Idiom of Kinship in Social Action among the Ndendeuli of Tanzania. Berkeley: University of California Press.

Hagedorn, John and Perry Macon
1988 People and Folks: Gangs, Crime, and the Underclass in a Rustbelt City. Chicago: Lake View Press.

Hall, Stuart and Tony Jefferson
1976 Resistance through Rituals. London: Hutchinson.

Hebdige, Dick
1979 Subculture: The Meaning of Style. London: Methuen.

Herzfeld, Michael
1988 Embarrassment as Pride: Narrative Resourcefulness and Strategies of Normativity among Cretan Animal-Thieves. Anthropological Linguistics 30(3/4): 319–344.

Hobsbawm, E. J.
1959 Primitive Rebels: Studies in Archaic Forms of Social Movement in the 19th and 20th Centuries. Manchester, UK: Manchester University Press.

Hopkins, Elizabeth
1973 The Politics of Crime: Aggression and Control in a Colonial Context. American Anthropologist 75(3): 7,312–7,342.

Hutchings, Suzi
1993 The Great Shoe Store Robbery. Oceania 63(4): 345–361.
Hyndman, David
1994 Ancestral Rain Forests and the Mountain of Gold: Indigenous Peoples and Mining in New Guinea. Boulder, CO: Westview Press.
Ianni, Francis A. J.
1974a Black Mafia: Ethnic Succession in Organized Crime. New York: Simon and Schuster.
1974b The Mafia Becomes an Equal Opportunity Employer: The Rise of the New Black Mafia. New York Magazine 7(4): 36–46.
Jacquemet, Marco
1992 Namechasers: How Belonging to a Community Became a Crime. American Ethnologist 14(4): 733–748.
Jenkins, Philip
1994 Using Murder: The Social Construction of Serial Homicide. New York: Aldine de Gruyter.
Kane, Stephanie and Theresa Mason
2001 AIDS and Criminal Justice. Annual Review of Anthropology 30: 457–79.
Keiser, R. Lincoln
1969 The Vice Lords: Warriors of the Streets. New York: Holt, Rinehart, and Winston.
Kelly, John Dunham
1990 Discourse about Sexuality and the End of Indenture in Fiji: The Making of Counter-Hegemonic Discourse. History and Anthropology 5: 19–61.
Kheng, Cheah Boon
1981 Social Banditry and Rural Crime in North Kedah. Journal of the Malaysian Branch of the Royal Asiatic Society 54(2): 98–130.
Koch, Klaus-Friedrich
1974 War and Peace in Jale'mo': The Management of Conflict in Highland New Guinea. Cambridge, MA: Harvard University Press.
Kozak, David L. and David I. Lopez
1999 Devil Sickness and Devil Songs—Tohono O'Odham Poetics. Washington, D.C.: Smithsonian Institution Press.
Kulick, Don
1993 Heroes from Hell: Representations of "Rascals" in a Papua New Guinean Village. Anthropology Today 9(3): 9–14.
Kuper, Leo
1981 Genocide: Its Political Use in the Twentieth Century. New Haven: Yale University Press.
Leach, E. R.
1970 Political Systems of Highland Burma. London: Athlone Press.
Lewis, Oscar
1966 La Vida: A Puerto Rican Family in the Culture of Poverty: San Juan and New York. New York: Random House.

Liebow, Elliot
1967   Tally's Corner: A Study of Negro Streetcorner Men. Boston: Little, Brown.
Malinowski, Bronislaw
1926   Crime and Custom in Savage Society. London: Routledge and Kegan Paul.
Marquez, Patricia
1999   The Street Is My Home: Youth and Violence in Caracas. Stanford: Stanford
        University Press.
Mayaram, Shail
1991   Criminality or Community? Alternative Constructions of the Mev Narrative
        of Darya Khan. Contributions to Indian Sociology 25(1): 57–84.
Mbembe, Achille
1992   The Banality of Power and the Aesthetics of Vulgarity in the Postcolony
        (Janet Roitman, trans.) Public Culture 4(2): 1–30.
McCone, Clyde R.
1966   Cultural Factors in Crime among the Dakota Indians. Plains Anthropologist
        11(32): 144–151.
McGillivray, Anne and Brenda Comaskey
1999.   Black Eyes All of the Time: Intimate Violence, Aboriginal Women, and the
        Justice System. Toronto: University of Toronto Press.
Merry, Sally Engle
1981   Urban Danger: Life in a Neighborhood of Strangers. Philadelphia: Temple
        University Press.
Merton, Robert K.
1949   Social Structure and Anomie. In Sociological Analysis. Logan Wilson and
        William L. Kold, eds. New York: Harcourt, Brace and Co.
Miller, Walter
1958   Lower Class Culture as a Generating Milieu of Gang Delinquency. Journal of
        Social Issues 14(3): 5–19.
Mishan, Ahrin and Rick Rothenberg
1994   Bui doi: Life like dust. Urban Nomad Productions. San Francisco: Cross-Cur-
        rent Media; National Asian American Telecommunications Association. Film.
Moore, Joan W.
1978   Homeboys: Gangs, Drugs, and Prison in the Barrios of Los Angeles. Philadel-
        phia: Temple University Press.
Moore, S. Falk
1978   Law as Process: An Anthropological Approach. London: Routledge and
        Kegan Paul.
Nader, Laura
1964   Talea and Juquila: A Comparison of Social Organization. University of Cali-
        fornia Publications in American Archaeology and Ethnology 48(3):
        195–296.
1990.   Harmony Ideology: Justice and Control in a Zapotec Mountain Village.
        Stanford, CA: Stanford University Press.

Nader, Laura and Harry F. Todd, eds.
1978   The Disputing Process: Law in Ten Societies. New York: Columbia University Press.

Nader, Laura and Philip C. Parnell
1982   Comparative Law and Enforcement: Preliterate Societies. Encyclopedia of Crime and Justice 1: 200–207.

Paredes, Anthony
1993   Capital Punishment in the USA. Anthropology Today 9(1): 16.

Parnell, Philip C.
1988   Escalating Disputes: Social Participation and Change in the Oaxacan Highlands. Tucson, AZ: University of Arizona Press.

Phillips, Michael R. and Thomas S. Inui
1986   The Interaction of Mental Illness, Criminal Behavior and Culture: Native Alaskan Mentally Ill Criminal Offenders. Culture, Medicine and Psychiatry 10: 123–149.

Phillips, Susan A.
1999   Wallbangin'—Graffiti and Gangs in L.A. Chicago: University of Chicago Press.

Pinney, Christopher
1989   Other Peoples' Bodies, Lives, Histories? Ethical Issues in the Use of a Photographic Archive. Journal of Museum Ethnography 1: 45–9, 57–69.

Pospisil, Leopold
1967   Legal Levels and the Multiplicity of Legal Systems in Human Societies. Journal of Conflict Resolution 11(1): 2–26.

Radcliffe-Brown, A. R.
1933   Primitive Law. In Encyclopedia of the Social Sciences, vol. 9. Pp. 202–206. Edwin R. A. Seligman and Alvin Johnson, et al., eds. New York: Macmillan.

Raphael, Jody
2000   Saving Bernice—Battered Women, Welfare, and Poverty. Boston: Northeastern University Press.

Regener, Susanne
1993   Verbrecherbilder: Fotoportrats der Polizei und Physiognomisierung des Kiminellen. Journal of European Ethnology 23(1): 67–85.

Ruffini, Julio
1978   Disputing over Livestock in Sardinia. In The Disputing Process: Law in Ten Societies. Laura Nader and Harry F. Todd, eds. Pp. 209–246. New York: Columbia University Press.

Said, Edward W.
1978   Orientalism. New York: Pantheon Books.

Scheper-Hughes, Nancy
1992   Death without Weeping: The Violence of Everyday Life in Brazil. Berkeley: University of California Press.

Schirmer, Jennifer
1998   The Guatemalan Military Project: A Violence Called Democracy. Philadel-
       phia: University of Pennsylvania Press.
Scully, Pamela
1989   Criminality and Conflict in Rural Stellenbosch, South Africa, 1870–1900.
       Journal of African History 30: 289–300.
Siegel, James T.
1998   A New Criminal Type in Jakarta: Counter-Revolution Today. Durham, NC:
       Duke University Press.
Sinha-Kerkhoff, Kathinka
1998   Juvenilization of Crime in Ranchi: Media's Creation of a Criminal Subcul-
       ture. The Eastern Anthropologist 51(2): 111–120.
Spedding, A. L.
1999   Dreams of Leaving: Life in the Feminine Penitentiary Centre, Miraflores, La
       Paz, Bolivia. Anthropology Today 15(2): 11–17.
Spradley, James
1970   You Owe Yourself a Drunk: An Ethnography of Urban Nomads. Boston: Lit-
       tle, Brown.
Sullivan, Mercer L.
1989   Getting Paid: Youth Crime and Work in the Inner City. Ithaca, NY: Cornell
       University Press.
Suttles, Gerald D.
1968   The Social Order of the Slum: Ethnicity and Territory in the Inner City.
       Chicago: University of Chicago Press.
Svaldi, David
1989   Sand Creek and the Rhetoric of Extermination: A Case Study in Indian-
       White Relations. Lanham, MD: University Press of America.
Thrasher, Frederic M.
1927   The Gang: A Study of 1,313 Gangs in Chicago. Chicago: University of
       Chicago Press.
Tolen, Rachel J.
1991   Colonizing and Transforming the Criminal Tribesman: The Salvation Army
       in British India. American Ethnologist 18: 106–25.
Tsing, Anna Lowenhaupt
1990   Monster Stories: Women Charged with Perinatal Endangerment. In Uncer-
       tain Terms: Negotiating Gender in American Culture. Faye Ginsburg and
       Anna Lowenhupt Tsing, eds. Pp. 282–299. Boston: Beacon Press.
Turner, Victor
1969   The Ritual Process: Structure and Anti-Structure. Chicago: Aldine.
Unger, Roberto Mangabeira
1976   Law in Modern Society: Toward a Criticism of Social Theory. New York: Free
       Press.

Vigil, James Diego
1988   Barrio Gangs: Street Life and Identity in Southern California. Austin: University of Texas Press.
Waterbury, John
1970   The Commander of the Faithful: The Moroccan Political Elite: A Study in Segmented Politics. New York: Columbia University Press.
Wedel, Janine
1999   Rigging the U.S.-Russia Relationship: Harvard, Chubais, and the Transidentity Game. Demokratizatsiya: The Journal of Post-Soviet Democratization 7(4): 469–500.
2000   Tainted Transactions: Harvard, the Chubais Clan and Russia's Ruin. The National Interest 59 (Spring): 23–34.
West, William Gordon
1974   Serious Thieves: Lower-Class Adolescent Males in a Short-Term Deviant Occupation. Ph.D. dissertation, Northwestern University.
Willigen, John Van and V. C. Channa
1991   Law, Custom, and Crimes against Women: The Problem of Dowry Death in India. Human Organization 50(4): 369–377.
Whyte, William Foote
1943   Street Corner Society: The Social Structure of an Italian Slum. Chicago: University of Chicago Press.
Wojcicka Sharff, Jagna
1997   King Kong on 4th Street: Families and the Violence of Poverty on the Lower East Side. Boulder, CO: Westview Press.
Yang, Anand A., ed.
1985   Crime and Criminality in British India. Tucson, AZ: University of Arizona Press.

CHAPTER 1

TRAVERSING THE Q'EQCHI' IMAGINARY

The Conjecture of Crime in Livingston, Guatemala

Hilary E. Kahn

Q'eq is a little black man who likes to roam during the night. Yes, but he looks like a dog. And he smokes a cigar. He begins to smoke. What he is going to rob is in the house, he will rob something there. He will come looking for things to steal. They say that this little black man eats eggs. When you pass nearby him, your hair will stand straight up. Yes, your hair stands up when he is nearby.

—From interview with Patrocinia, May 14, 1996

I REMEMBER THE DAY A YOUNG GARIFUNA MAN committed suicide in Livingston. He was a popular youth and the funeral was too emotional to be translated into ethnographic text. My fieldnotes simply read, "so much wailing, crying, agony, fainting." The town mourned this tragedy. The next day I spoke with Q'eqchi' people who told me that the young man had raped a woman and that the police were after him. To avoid going to jail, he killed himself. This was not true. For a moment I was angry and repulsed. After experiencing the deep emotional pain at the funeral the day before, I then saw the cold manner in which my Q'eqchi' friends defined this young man as another criminal, *un ladron*. The way in which they explained this incident demonstrated their cultural construction of crime. It revealed how defining crime is a process that blocks compassion. Although sometimes the result of this process is an utterly arbitrary explanation and assignment of blame, crime is determined within the same cultural and historical webs that identify selves and others. The sharp juxtaposition spurred me to investigate just how these cultural beliefs surface.

Q'eqchi' Mayan migrants label themselves and others through a network of exchanges that span the local and global. Historical, social, and mythical connections—forged in extrapersonal structures and internalized cultural models—are also the intimate and institutional relations that filter contemporary Q'eqchi' perceptions of wrongdoing. They likewise are recreated in ethnographic experiences, theories, methods, and texts. Myth and history, the local and global, and the tangible and intangible merge throughout the process of conducting fieldwork, in the text of this essay, and in the practice of Q'eqchi' people defining criminals and their crimes in the Caribbean town of Livingston, Guatemala.

LIVINGSTON: THE (UN)MEETING OF HISTORIES

The Q'eqchi' people are originally from Alta Verapaz, a highland region of Guatemala that is considered the cradle of their culture. They began their trek toward the Caribbean over 100 years ago, although some have only recently left the mountains of Verapaz in search of work, land, and freedom. Many come to Livingston, a small lowland fishing town on the Caribbean coast that has no roads leading to it, no autochthonous indigenous Mayan population, and no town square. More Caribbean than Latin American, this is not typical Guatemala. Rather, it is a sundry town full of diverse ethnicities, histories, and economies.

Livingston is on the northwest side of the bank of the Río Dulce where the river opens up to meet the Bay of Amatique. Founded by Marcos Sanchez Díaz in 1802, Garifuna people are the original settlers of this town. Arriving in the Americas on an African slave ship, the Garifuna came to Guatemala after living free on the island of Saint Vincent for centuries, there mixing with Carib islanders.[1] Today a small fishing town with a growing economy based on tourism, Livingston was once the primary Atlantic port of Guatemala. Both Ferrapazco,[2] a German-owned export and shipping company, and the United Fruit Company loaded up their boats here, steering them to the United States and Europe full of coffee, cacao, bananas, and wood. By the time the Guatemalan government expropriated the property and assets of

German companies during World War II, including those of Ferrapazco, the emphasis on export had already shifted from Livingston to Puerto Barrios. By the 1960s, many of the Garifuna people began immigrating to the United States, a practice that continues steadfastly today. The economy of the Garifuna who do not migrate is firmly dependent on those who do.

Q'eqchi' people began moving into Livingston's surrounding bush some 50 years ago. Only within the past 10 to 15 years have many of them begun living within the town itself. There they come upon many novel situations. Barefoot hippie tourists, the ocean, black people, and reggae music are encountered for the first time. Both Q'eqchi' men and women tell me particularly how frightened they were when they first saw black people. Calling them *q'eq* (black), the Q'eqchi' were fearful that the Garifuna would hit, grab, or rob them. Although most Q'eqchi' people tell me that they are no longer as afraid as when they first arrived, fear is still pervasive. My Q'eqchi' friends would continuously show concern for my safety when I walked home alone at night, warning me about the potentially dangerous *morenos*.[3]

The misunderstanding of the Garifuna by the Q'eqchi' dramatically reveals the internalized model of perception, fortified through centuries of relationships with outsiders, that governs the ways in which the Q'eqchi' Mayans identify transgression. Foreign landowners, white ethnographers, and Ladinos (people of mixed indigenous and European ancestry) easily fit into this Q'eqchi' taxonomic framework. We are conceived in typical terms of foreignness—ambivalently—as benevolent providers and malevolent thieves. The Garifuna, however, seem to be located between Q'eqchi' categories of otherness. They are not white, rich, land-owning overseers. Yet, through migration to the United States, they are thoroughly entrenched in transnational connections. They are therefore passersby, not so unlike the tourists, guerrillas, and mythical beings who also traverse the Q'eqchi' imaginary. Because the Q'eqchi' cannot easily classify the Garifuna, they ultimately filter their perception of them through images of the Q'eq, the mythical black trickster whose narratives embody the practice and perception of social deviance.

THE ORIGINAL TRANSGRESSORS

Spaniards were the first criminals. Q'eqchi' people tell me that Guatemala was once rich, full of gold, money, and treasures. Besides stealing monetary wealth, the Spanish stole women. They carried their loot across the ocean, which is why Spain is rich and Guatemala is poor. This poignant allegory of colonialism and development represents the first infringement the Q'eqchi' people of Alta Verapaz confronted. During the sixteenth century that followed, Bartolome de Las Casas, the so-called defender of the Indians, experimented with *los Indios* of Verapaz to justify non-violent conversion to Christianity and the Crown. While Las Casas did save the indigenous of Verapaz from the brutal treatment experienced in other parts of Guatemala, they were still forcibly gathered, taxed, and baptized. Fleeing the *reducciones*[4] for homes in surrounding mountains, Q'eqchi' migration was an innovative form of resistance.

After Guatemala's independence from the crown in 1821, Alta Verapaz became a popular spot for Ladino and foreign agricultural development. By the turn of the century, this region had become a veritable German colony with coffee its staff of life. In order to assist the Germans in their business ventures, the Guatemalan government offered incentives such as tax exemptions and cheap land and labor. They gave out coffee seeds, built roads, and opened up communication by installing a telegraph system. The government perceived the Mayans as a labor supply, and they used coercive and hostile means to assure them as such. Debt-peonage and the *mandamiento*[5] system provided the needed work force by demanding labor from "vagrant" individuals who did not own sufficient land to provide an income. Because most Q'eqchi' Mayans were unable to raise the capital to buy their own ancestral lands, they became bound by debt to both the government and foreign landowners. They watched helplessly as Germans and Ladinos bought up much of the fertile land in Alta Verapaz.[6]

Q'eqchi' people migrated away from Alta Verapaz, beginning the process that would eventually land some of them in Livingston. Others stayed to live and work on the German-owned coffee *fincas* (plantations). On these fincas, Q'eqchi' people tell me, the Q'eq was born:

> My father told me that when he lived in Alta Verapaz, he worked with a finquero, and this finquero owned some two or three Q'eq. He says that they are like dogs. They guard the coffee plantations so that robbers won't enter, so he tells me. Many Q'eq existed in Alta Verapaz. I'm not sure if they still exist today, but people say they do. Sometimes at night when someone hears whistling from the bush, they say it is the Q'eq passing by. He is an animal who walks but fast, fast, fast! They say he is as fast as a car, but he is an animal (From interview with Martín, June 20, 1996).

Q'eq means black in the Q'eqchi' language, although people would laugh when I asked if there was any sort of linguistic affiliation between this term and their own Q'eqchi' identity. For them, Q'eq indicates other. It is a little black man that is part-human and part-animal. Most people describe him as part-dog. Conceived when a man had sex with a cow, this little black animal is the guardian of the fincas. Q'eq are owned by the German *finqueros,* some of whom possess as many as five or six of these animals. Usually described as smoking a big fat cigar, they are black, furry, and short, and their elbows and knees are placed backwards.[7] When people enter the finca to steal a cow or coffee plants, the Q'eq grabs them and takes them to the owner's house for sentencing. Guarding cattle like a dutiful son, Q'eq protects foreign owners and his bovine parents, both of which are European imports symbolizing foreign commerce.

How this myth arose is not a simple question. Most scholars assume that the Q'eq were mythologized as a result of the blacks brought over after the Conquest (Bricker 1981; Cabarrus 1979). African slaves were indeed brought to the New World as early as the sixteenth century, and the Germans were known to use blacks from Jamaica on their fincas. However, although the myth was strengthened with the arrival of the Germans, I suggest that this myth may well have a pre-Columbian origin. God M is a Post-Classic (A.D. 900–A.D.1500) deity who was black. He most likely became *Ek Chuah,* the contact period deity from the Yucatan, who was also a black god with special

1.1. God L, a Classic period precursor of God M, whom the Q'eqchi' conflated with black slaves on German *fincas* (redrawn by the author from Taube [1992], which is after Robicseck [1978]).

significance for cacao growers. Both God M and Ek Chuah (*ek* is *black* in Yucatec Mayan) were known as merchant gods who were worshipped by traders and travelers (Taube 1992). The origin of Q'eq is with these deities who were eventually conflated with the few black slaves used by the Germans. Still associated with business, money, agriculture, and foreign travel, the Q'eq is not simply a colonial invention. Once worshipped by the ancient Maya, through colonialism and capitalism Q'eq has now been inverted into a negative image. However, due to its association with foreign owners, this disrespectful little entity ambivalently deems respect and reverence.

Q'eq reside on the fincas of Alta Verapaz, although people tell me that they come to Livingston to steal. Slipping out of their fincas at night, they roam the countryside in search of things to rob and people to scare. When all the dogs bark in the middle of the night or when one hears an unexplained whistle, a Q'eq is said to be passing nearby. They rob anything they can find—from money to shoes to chickens—although their favorite take is eggs, which they eat by the dozen. They love to consume. They bring the stolen articles to their *dueño* (owner) at the finca. Stealing from the poor and carrying their loot to the rich, they are an inversion of Robin Hood (Cabarrús 1979).

When you see the Q'eq, you will not be able to move. You cannot run or walk, and your hair will stand straight up. You will want to scream, but you will find yourself mute. Patrocinia, a Q'eqchi' woman, told me about the day her brother saw a Q'eq in El Estor. The Q'eq was standing under a tree, smoking a cigar. Wondering who this strange person was, her brother picked up a rock to protect himself. As he got closer, he became more frightened with every step. He decided that it would be better if he left, but he was frozen with fear. The Q'eq did not rob or injure her brother. He only terrified him. Accordingly, Patrocinia tells me that the dueños of the fincas have these Q'eq *solo para sustarnos* (only to scare us).

Single women, such as I was in the field, must be particularly careful to shut their windows at night to prevent the Q'eq from entering. Although never explicitly stated, people would infer the voracious sexual appetite of the Q'eq. Consider the overt symbolism of his favorite food—eggs—and the act of bestiality through which he was conceived. By consuming symbols of fertility, this mythical creature is highly sexualized. Many Q'eqchi' girlfriends would ask me if I was scared of the animal spirits who could enter my house and come touch me while I was sleeping. Truthfully, I did become frightened. Not only was I somewhat infected with my friends' fear of nighttime spirits, but I also realized how the mythical entities were metaphors for something real that indeed merited my fear. I weekly sprinkled a mixture of rue and garlic around my house in order to keep away spirits, snakes, and criminals.

FEAR AND FOREIGNERS

According to the Q'eqchi', once a criminal always a criminal. Perhaps because justice is so rarely experienced in Guatemala, rehabilitation is not an option. Criminality is viewed as an essential feature that cannot be changed. According to the Q'eqchi', many of the Garifuna were stricken with an *enfermedad* (sickness) that made them not want to work. With no available cure, Q'eqchi' people would tell me that the morenos' only option was to steal. Clearly understanding the correlation between unemployment and crime, my friend Chico told me, "violence comes with misery when there is no work. When there is work, there is no more violence. I believe the morenos worked with the company. Yes, there was a company with whom the morenos worked hard." Chico is speaking of the United Fruit Company and the fact of the matter is that, yes, the Garifuna once lived in a town with a booming economy; however, in the 1940s, Livingston hit hard times, and in the 1960s immigration to the States began. Garifuna relocation to Los Angeles and New York has an incredible impact on Livingston, and on the Q'eqchi' perception of their neighbors. Every Garifuna person I know has at least one relative, whether it be a sister, a mother, or an uncle, currently working in the States. While earning dollars up north, these immigrants send money to their relatives in Livingston. Replacing one foreign enterprise with another, Western Union has become the supplier of much of Livingston's economy.

Those Garifuna who remain at home in Livingston find themselves in a position in which jobs are scarce. They may find work in the tourist industry as motor boat drivers or assistants, tour guides, bartenders, servers, hotel procurers, or street musicians. Garifuna people who find work outside tourism are civil servants, educators, health professionals, carpenters, secretaries, and town employees. An increasing number of Garifuna people buy businesses—discos, bars, restaurants, clothing and video stores—often with dollars earned in the United States.

In contrast, wired money, sex, drugs, and foreigners are inseparable signifiers in the Q'eqchi' perception of crime in Livingston. Q'eqchi' people tell me that foreigners bring drugs and diseases to the area. They tell me that tourists want to have sex with locals, and locals want to find themselves a foreigner. They say that checks from the States make the Garifuna not want to work but rather sit around and wait for the next installment. Q'eqchi' say that when their money runs out, the Garifuna steal from their favorite marks—the tourists. Clearly, the Q'eqchi' understand foreign people and money as intricately involved in the illness that supposedly makes the Garifuna steal, do drugs, and rape. Even so, Q'eqchi' people also desire American dollars. Like Western clothing, foreign currency is sought after at the same time that it's equally held responsible for debasing and sexualizing the individuals receiving and displaying it. And similar to the stories I heard of foreign treasures scaring people or leading to doom when unearthed and sold, Western Union supplies a symbolic and real currency that lends itself to a construction of fear, particularly because the Q'eqchi' usually do not benefit from that currency in a positive way.

Gossip circulated when Q'eqchi' women saw me speaking with any Garifuna man. They would ask me if I was keeping him company at night or if we were making babies. When I denied these accusations, they would persist, joking about the sexual relations I had with every moreno I was seen talking with in the street. Q'eqchi' people explained to me that drugs made the morenos sexual, lazy, and deviant. Q'eqchi' people referred to the getting high process as *fumando sus puros* (smoking their cigars), which made them *waxiru* (crazy):

> I saw some *gringas,* I don't know where, on the beach, on the beach at Cayo San Jose. One of them was, you know, together with . . . I saw this, I saw it with my own eyes. I saw her and she didn't have anything, no children, nothing. But when I saw her the next time, she had children, but little black ones. Can you believe it? From a gringa came little black ones! So things go now. . . . You know, it's the same with German blood, as it is now mixed, as I told you (from interview with Don Alberto, February 29, 1996).

In various meanings of the word, foreigners have always come to screw the Q'eqchi' Mayans. After the Spaniards, it was the Germans. Neither group of intruders are said to have brought women, so men had to find female companions to fulfill all their domestic needs. Many Germans took Q'eqchi' women as mistresses, and this is why, it is explained to me, numerous Q'eqchi' people have light hair like mine. This is why when I wore my *uk* and *poot* (traditional skirt and embroidered cotton blouse) some Q'eqchi' would ask if I were a visiting relative from Cobán, the largest city of Alta Verapaz where it is said many pale-skinned blonde Q'eqchi' reside. Cobán was the pulse of the German coffee industry at the turn of the century.

While many Q'eqchi' talk about the Germans with admiration and respect, there is an ambiguous *fear of the foreign.* Foreign authorities and institutions, such as plantation owners, while simultaneously providing, protecting, and feeding their Q'eqchi' workers and subjects, also appropriate, enslave, beat, and scare them. So extreme is this fear that some Q'eqchi' people ask me if the disease-carrying North Americans eat children. Undoubtedly, this *fear of the foreign* creates the backdrop for the prevailing fear of children-stealing gringos. Numerous times I heard rumors about groups of gringos that were wandering through Livingston in search of children to bring back to the United States. In 1994, this cultural belief made national headlines when Poqomchi' Mayans were reported to have severely beaten and raped an Alaskan environmentalist accused of child-stealing in Alta Verapaz. Unlike the U.S. media that represented this cultural fear as irrational and ignorant, I understand this panic as originating from a long history of foreign intervention, rape, murder, and economic control. Why shouldn't Q'eqchi' people assume that North Americans, as cultural and economic ambassadors, come to Guatemala with the same evil intentions? Tragically, although the North American victim had only good intentions, she was ignorant of, or did not identify with, the evil lurking within the structures of colonialism, global consumption, and third world poverty.

But what about the Garifuna in Livingston? Where do they fit into the Q'eqchi' mythic-history? They are not really foreigners, nor are they typical Guatemalans. They tend to pay their nationalistic allegiance to the Garifuna communities that reside along the coasts of Honduras, Guatemala, and Belize. Due to migration, Q'eqchi' perceive them as pseudo–North Americans, even though they conflict with the traditional image of the United States gringo—white, wealthy, and living in guarded mansions. Unlike Germans, the Garifuna are not foreign owners who require respect and fear due to historical relationships of unequal power. In fact, like the Q'eqchi', they too are victims of colonialism, capitalism, and various other foreign forms of institutional intrusion. Moreover, the Garifuna also fear and dislike Ladinos (perhaps even more than the Q'eqchi' Maya). Interestingly, this particular oppositional disdain is picked up by Mayan cultural organizations who come to Livingston to discuss issues with the Garifuna people. Now officially defined in the Peace Accords, the Garifuna, like the Mayan people of Guatemala, are indigenous. Equally, they have rights to an indigenous identity and culture. Oddly, most of the Mayan revitalization organizations that come to town have no knowledge that some 500 Q'eqchi' individuals live in Livingston. Based upon an inappropriate assumption about links between ethnicity and place, Mayan organizations ultimately label the Garifuna as more indigenous than the local Q'eqchi'.

In a similar process of misinterpretation, the local Q'eqchi' came to understand their Garifuna neighbors as criminals by incorporating contemporary relations into historic social categories. The Q'eqchi' people came to Livingston with histories and ideas of others constructed through relationships of power.[8] For example, because they had plenty of experience dealing with foreigners and Ladinos, my role as foreign anthropologist was easily understood. First I was feared, and then fright mixed with respect. Blonde and blue-eyed, I offered concern, money, medicine, and friendship. However, with my camera, tape recorder, notebook, and computer, I stole images, histories, and cultural riches that I then carried to foreign places. Exporting ethnographic knowledge rather than coffee, I was easily and ambiguously understood as a metaphoric German finca owner.

On the other hand, the Q'eqchi' do not envision the Garifuna as foreign owners demanding payment or labor. They are imagined as indigenous, yet also entrenched in a global economy. Because many of them live in the United States and only come to Livingston for holidays, they are perceived as passersby. Two modes of production—a Q'eqchi' one that involves local community exchange with owners and the land and a Garifuna one whose community is stretched and hidden through transnational networks of migration—create further difference and confusion. And then there is their skin color, which leads the Q'eqchi' to utilize an appropriate reference point of conflation—the mythical black Q'eq.[9] Because Garifuna apparently fail to stress reciprocal exchange between land, individuals, and owners, the Q'eqchi' identify the Garifuna as different, immoral, and natural transgressors. Valuing foreign currency, appearing idle or only passing by, the mythical Q'eq is manifested in the (mis)perception of the Garifuna as selfish and disrespectful.

My girlfriend told me that the morenos are mean at the pump where she goes to get water. I asked her why, and she replied, "God knows, because they are morenos maybe, if you go to get ahead in the line, they'll hit you. They are bad." Numerous other people would tell me that the Garifuna *no tienen respeto* (they do not have any respect). Q'eqchi' people commonly utilize the issue of respect to explain differences between people. Q'eqchi' people tell me that their ancestors used to respect each other, the land, and God. Complaining that people do not pay respect anymore, they tell me that the worst offenders are the Garifuna. Nonetheless, Q'eqchi' people do not spare anyone who demonstrates lack of respect. This defining characteristic is also used to differentiate among the Q'eqchi' themselves, particularly with regard to religion. Elderly people complain that the youth do not respect God anymore, and everyone tends to criticize the general lack of morality in Livingston. Crime, let alone poor crop yields, bad weather, rodents, insect infestations, illness, and death are all results of this decaying moral fabric.

If everyone is at fault, why are the Garifuna labeled as the worst offenders? Why are they perceived as any more disrespectful than the next? The answer is most likely in their particular economy and associated system of values. Q'eqchi' people tell me that the Garifuna used to work the land and grow cassava, yucca, plantains, and bananas. However, they are no longer agriculturists, and Q'eqchi' say that today their neighbors only wait for checks from Western Union and hustle tourists. No longer connected to the land but to the United States, Q'eqchi' people accurately understand Garifuna economy as a circulation of foreign currency. Whether dollars or international tourists, what the Garifuna value is foreign and, according to the Q'eqchi', self-serving.

Contradicting this cultural construction, the Garifuna in Livingston in fact use their remittances to buy food, build houses, purchase medicine, educate children, finance businesses, and pay debts. Plus, like increasing numbers of indigenous people globally, the Garifuna also maintain an active traditional culture through foreign money. Secured in Livingston, the New York and Los Angeles Garifuna have a place they call home. Through wired money, they maintain their future. When Garifuna people come yearly to Livingston to participate in festivals and rituals, most know that one day they will return for good. However misunderstood by the Q'eqchi', the Garifuna maintain their local community through global economic networks. Immigration displaces human bodies and, through the return of hard currency, balances the exchange between what was once considered a consuming core and a drained periphery.

The Q'eqchi' lack of compassion and their inaccurate reading of the Garifuna may also be related to the visibility of the hustler street scene in Livingston, where mostly young men tend to make false promises to naive foreign tourists. Apparent also are the Garifuna living in the States who come through town wearing shiny gold necklaces and clean Nike sneakers, carrying loud boom boxes, and bringing similar foreign luxury items to their loved ones. What the Q'eqchi' see then are the fruits, not the labor. Nor do they see how the Garifuna culture and morality is so similar to their own. They see only difference. They fail to understand how the Garifuna may have their own version of moral exchanges that they use to define themselves and their community. They find only a living myth.

PRACTICING MORALITY AND BLOCKED COMPASSION

Q'eqchi' identities are created and maintained through moral and economic exchanges among owners, individuals, institutions, communities, and the land. Many Q'eqchi' men today work as wage laborers in town, although the majority still maintain corn *milpas* (fields) in the surrounding bush. Sowing or harvesting these fields is a community activity, one that is considered more than a profane economic experience. Through sexual abstinence, food restrictions, prayer and ritual, milpa growers pay their due respect to Tzuultaq'a, the bi-gendered mountain spirit who offers permission to dig the earth. This reciprocal reverential exchange is necessary any time an individual or community wants to utilize those natural resources possessed by the mountain spirit, whether it is breaking ground for a new home, digging for a well, hunting wild animals, or planting corn. However, please do not misinterpret this land-focused framework as revealing an essential circulation of Qeqchi' culture through spiritual connections with their geography. Although agricultural subsistence does continue to motivate contemporary practices of morality, this physiocratic bond is predominantly a metaphor for stable and cohesive identities in the midst of motion.

The Q'eqchi' believe that everything has an owner, whether it be possessed legally, spiritually, or physically. Q'eqchi' morality enforces that people pay respect to these owners. Thus, when people want to sow their fields, build a house, or cut down a tree, they must ask permission from Tzuultaq'a, the true owner of the land, through ritual and sacrifice. This is someone's debt *(c'as)*. When someone offers the mountain spirit copal pom, a chicken, or candles, s/he repays the individual by providing for them. S/he gives health to children, safety to travelers, and good yields to farmers. When death or blight occurs, it is the individual's fault for not believing strongly enough or for thinking or acting immorally. If someone acts inappropriately or disrespectfully, Tzuultaq'a sends snakes out to scare or kill the violator. In accordance with the typical recipients of Q'eqchi' debt, Tzuultaq'a is often depicted as a foreigner, as an old white man or woman who lives within the mountains. Inside the home are vast amounts of gold, silver, food, and animals that s/he shares with the most reverent followers. These, though, are not stolen treasures. These things were offered to the mountain spirit in exchange for granting permission to sow, eat, build houses, and to live. Therefore, although Tzuultaq'a too is foreign and solitary, s/he is morally in opposition to Q'eq because s/he embodies and adheres to the rules of Q'eqchi' morality.

Q'eqchi' morality is built upon centuries of ritual and everyday exchange between individuals, gods, land, and owners. Through pre-Columbian means of land distribution, colonialism, Catholicism, debt-peonage, and capitalism, the Q'eqchi' people have continuously been involved in social, sacred, and economic situations that reinforce their lack of ownership. Even when the Q'eqchi' assume they are in possession, deities, governments, churches, Germans, Ladinos, international nickel companies, or various other foreign institutions ultimately claim the status of true owners. Through work, ritual, tax, debt, or loss of land, the Q'eqchi' end up offering sacrifice to these owners. Involved in a cycle of debt, the Q'eqchi' are securely bound to a transcendental owner, whether it be God, Tzuultaq'a, the government, or a German patron. For more than 500 years, economics, religion, and politics have worked together to create this internalized framework.[10]

Therefore, when Q'eqchi' people arrive in Livingston for the first time, they are equipped with this structural arrangement. Because the Garifuna people are between categories—neither fully foreign, white, or indigenous— the Q'eqchi', in responding to the Garifuna, draw on the mythical Q'eq, which emerges from historically reinforced ideas of morality. Q'eq originates from within the Q'eqchi' moral model simply by his opposition to the system. Physically displaying his societal deviance in his backward elbows and knees, Q'eq represents the foreign, mischievous, sexual, and discordant. Symbolizing the unknown and adversity, Q'eq is bad weather, poor crops, and blight. Owned by German finqueros, who themselves are metaphorical criminals of land and labor, Q'eq's sole purpose is to scare, steal, secure the finca, and eat eggs. Q'eq is also a solitary figure, which opposes Q'eqchi' ideas of community. Consuming symbols of reproduction, acting individually, and propelled by foreign interests, the Q'eq is the epitome of disrespect and nonproductivity.

When I asked the Q'eqchi' if the Garifuna and the Q'eq were the same, they told me that they were different but similar. Although Q'eqchi refer to the morenos as Q'eq, Garifuna people cannot turn themselves into animals, so they are not *really* Q'eq. However, because they are both perceived as cigar-smoking black criminals, Q'eqchi' reply that they may be related. Although clearly differentiating between the Garifuna and Q'eq, the Q'eqchi' do seem to apply a manifestation of this mythical and historical figure to their understanding of their black neighbors. This is why they create such shocking and distorted images. They tell me their Garifuna neighbors are rapists, drug addicts, and criminals. They are supposedly violent, mean, inconsiderate, and kill dogs just for fun. Being non-producers, they are simply wasteful consumers. Frivolously, they participate in wanton rituals where food is thrown out for the dead ancestors.

A friend once told me about the Garifuna ritual *dugu* in which, he said, the participants waste food by throwing it into the water. I asked if the Q'eqchi' did not do a similar thing during their Todos Santos ritual on the first day of November. Don't Q'eqchi' people leave food out for ancestors on this day? He was quick to respond that yes, they do, but they would never throw the food away. They leave it out for *los muertos* (the dead) during the day, but then people later return to eat it. *Ellos lo tiren, nosotros lo recogemos* (they throw it out, we gather it up). Garifuna extravagance is supposedly demonstrated through other rituals, such as during Ash Wednesday when they squander the day throwing flour at Ladinos and tourists. Often complaining about the nonproductivity of this ritual, Q'eqchi' people prefer to spend this day working. For the Q'eqchi', ritual is always a debt or payment in exchange for something else.

Garifuna rituals are misperceived as wasteful, even though remaining food from the dugu is customarily sacrificed to ancestors or redistributed among the poor and needy (Jenkins 1998:162). They are misconstrued as passersby, as transnational sightseers—coming through Livingston for brief visits, avoiding connections with the land and the Q'eqchi' community, and displaying their inclination toward consumption of foreign goods. They are imagined as cigar-smoking thieves who pilfer for foreign interests. Rather than bringing the mythical loot to foreign landowners, however, the Garifuna earn dollars, buy American products, strengthen the U.S. economy, and mingle with pale-skinned tourists. Blocking compassion, negating what to some are obvious similarities, and explaining differences through a filter of historical and social exchanges, the Q'eqchi' label the Garifuna criminals and explain their transgressions.

Digging deeper, one encounters even more irony in the way the Q'eqchi' categorize their Garifuna neighbors as criminally inclined foreigners. Nancie Gonzalez (1988:110) and numerous locals in Livingston believe that the international immigration that precipitated Garifuna participation in a transnational network was actually a result of the migration of the Q'eqchi'. Because the Q'eqchi' began to build homes and plant corn on much of the land that surrounds the town, after the mid-century decline of the local economy the Garifuna could not return to agriculture. They had few economic options other than to migrate to the United States. Thus, it is believed by many that Garifuna migration was instigated and perpetuated by Q'eqchi' occupation of local land. If this is so, then the irony emerges from the fact that the Q'eqchi' create not only their own criminals but also their own foreigners. They take what the Garifuna reject and then renounce the alternatives they choose. They manifest immorality in the bodies of those engaged in global practices and morality in their own commitment to a localized mode of consumption and production. Q'eqchi' people position themselves as the producers, the consumers, and those consumed.

INSTITUTIONALIZING IMMORALITY:
TODAY'S CRIMINALS

While I was in Livingston, there was a young Garifuna man who did not play by the rules. Everyone knew he was trouble, robbing homes, businesses, and tourists on a daily basis. He was murdered on the Guatemalan presidential inauguration night. We knew it was bound to happen, for this is how justice works in Livingston and in the rest of the country where lynchings are becoming extremely common. The new president's modernizing right-wing party, National Advancement Party (PAN), beat the Guatemalan Republican Front (FRG), an extreme right party with old ties to the military. I was glad that Alvaro Arzú won the presidency, but many of my Q'eqchi' friends were not. They had listened to FRG's charismatic leader Rios Montt's political promises of bringing Guatemala back to the Guatemalans. More importantly, the part of the rhetoric that the Q'eqchi' people emphasized to me was FRG's vehement stand against crime and foreign intervention. Because PAN won, they were afraid that foreign countries were going to invade Guatemala through military and economic means. Moreover, they now feared the uprising of criminals. For them, increased crime accompanies foreign intervention.

The young Garifuna rule-breaker is an example. I heard from a Q'eqchi' friend that he was drunk on the streets earlier on inauguration evening. He was supposedly dancing, singing, and jumping around in ecstatic glee because PAN had won. This meant he was now free to continue his crime spree. He was screaming and singing "Gracias a Dios que ganó PAN. Sí ganó FRG, iban a quitar todos los *ladrones!*" [Thank god PAN won. If FRG had won, they would have kicked out all the criminals!]. Later that night he was murdered. I question if this actually occurred, because the young man was said by others to be hiding out. The Q'eqchi' version of events, however, does demonstrate the way in which state politics intersects with the Q'eqchi' broad understanding of crime. Far from apolitical, their morality is firmly grounded in and influenced by contemporary national and international politics. Yet, their interpretation of the state is equally construed through their own perception of morality and mythic-historical structures. Rather than having an invisible authority that categorizes and castigates its subjects, the Q'eqchi' also construct what is the state—a criminalizing institution with foreign interests.

While unraveling the histories, economies, and politics that have created contemporary Q'eqchi' perception of crime and authority, there is an additional political figure that should be addressed—the *guerrilla*. Although most of the Q'eqchi' people now residing in Livingston did not directly witness the battles of the Guatemalan civil war,[11] they were influenced by its practices. To the Q'eqchi', guerrilla does not necessarily signify the political fighter of the Guatemalan National Revolutionary Unit (URNG); rather, guerrilla simply means criminal—ladron. Being a liberal social scientist, I was shocked the first time I realized this conflation of guerrillas with ladrones. Don't the Q'eqchi' realize that these soldiers are fighting for them? The fact of the matter is, yes, they were aware that the URNG supported indigenous rights and land reform. However, they nonetheless would whisper to me that there were guerrillas in the area and that they were very dangerous. In fact, one friend of mine told me that they had killed one of the guerrillas on inauguration night in Livingston, although he was not the bad kind. This guerrilla was the kind that only stole things such as radios. Someone else told me about a group of guerrillas taking over a bus in Semux. They entered the bus and made everyone give them 20 quetzals. My friend was upset because the man on the bus had to give up his last 20 quetzals, leaving him penniless and without any means of returning home. What kind of help was that, he wanted to know?

My friend Javier tried to explain. He told me that in many villages guerrillas enter and make people give them food rather than money, forcing them to kill cows, turkeys, and chickens. While consuming the village sustenance, the guerrillas tell locals that they will fight for them and take care of them. But have they? How has life changed? In the eyes of many, the guerrillas simply come in and take what is not theirs. Q'eqchi' people are quite poor and the loss of a cow or even a turkey can financially hurt a family. To the Q'eqchi', guerrillas most definitely are criminals, a sentiment similar to one revealed in war-torn areas of Guatemala, where Mayan people find themselves caught between two armies (Stoll 1993). Irregardless of the ethnic and philosophic connection the Q'eqchi' may have with the rebels, they are nonetheless envisioned—like the mythical Q'eq—as moving covertly through the forest, consuming without producing and taking without permission.

If neither the state nor the rebels can protect you, who will? My Protestant friends tell me that Jesus is available. Because God's children are protected from all evils, they cannot be bitten by a snake or killed by a guerrilla. I heard a story about a pastor who was preaching in a guerrilla-controlled area. Labeling him a troublemaker, the guerrillas decided that this pastor had to be killed, and one night their boss went to shoot him. But when the guerrilla boss tried to take the gun out of his holster, his hand began to tremble and he was not able to lift his arm. He was frozen. The pastor was saved by God.

Unfortunately, not everyone is so intimate with Jesus. Even after the signing of the Peace Accords, Guatemala continues to be a violent country. It continues to be the number one car-jacking country in the world. Over 400 people were kidnapped in the first three months of 1997 (from *Prensa Libre* April 5, 1997), and lynching is a routine form of justice. Everyone tells me that in order to reduce crime, the government must find work for the people. The Peace Accords have promised employment and development projects, but the people in Livingston have yet to see any. Millions of dollars in foreign aid were sent to help Guatemala, but most people believe that the government is keeping this money to buy more land. One man told me that this international support goes directly to the military and politicians who use it to buy mansions in the United States. "They want to own everything," he remarks about the government. Still hidden and out of reach, foreign currency pours into Guatemala; secured not in mountains but in banks, foreign interest-minded owners keep it in their own hands. "The authorities don't help the people who need it; it is all for themselves." Q'eqchi' people complain about election promises that are never fulfilled and politicians who forget their platforms upon entering office. Q'eqchi' in Livingston say that even the president does not take the indigenous people into consideration; he cares only about his friends. Not completing their end of the bargain, these political authorities of the state are emerging as the latest criminals. Things seem to be getting worse.

Conditions are becoming so bad that people tell me the Q'eqchi' are beginning to steal. In a local village, Q'eqchi' *bandidos* raped a girl and stole thousands of quetzals from her mother. This same group of men were stealing cattle from an Italian-owned finca. They are considered extremely dangerous. Local finca owners now pay their workers in checks rather than cash because they are afraid to carry large amounts of money in the bush. Because I heard various versions of the crimes, I do not know if the bandidos were really Q'eqchi'. What I know was that, previously, such a possibility had been unheard of. When I began fieldwork in 1994, Q'eqchi' people blamed only the Garifuna for the crime and violence in Livingston. Three years and a Peace Accord later, Q'eqchi' are accusing each other.

What happened in the few years to bring about this switch is something about which I can only speculate. Both the accusations and the criminal act may be influenced by media reports of the violent crime wave that has occurred in Guatemala since the signing of the Peace Accords. Accordingly, everyone in Livingston talks about the inordinate crime occurring in the country. Lack of faith in the authority of the state grows daily. Doubting their government's ability to help, the Q'eqchi' voice a concern also raised by the Garifuna, thus unifying these people against the state. Mayan Cultural Revitalization also effects consensus. Coming to Livingston and proclaiming the equality of the Garifuna and the Q'eqchi', Mayan organizations emphasize similarities over differences. A few people tell me that these organizations are beneficial because they teach the Garifuna that the Q'eqchi' are human while demonstrating the same about the Garifuna for the Q'eqchi'. Under the law of the state, they are all indigenous people with rights to their own ethnic identities. Catholicism also helps to unify the two groups by promoting cultural equality.

The changing economy definitely plays a role. Slowly moving away from agriculture and working in wage labor and within the tourist industry, the Q'eqchi' value system is also adapting. The structure of their morality is transforming as it creates new myths and reacts to new histories, economies, and people. Still valuing a transcendental exchange with land and the community, Q'eqchi' are also beginning to deal with national and foreign currencies, tourists, and luxury items. My older friends complain about the youth who no longer tend to their corn milpas nor ask permission to cut down a tree. These days, the young are more concerned about buying clothes, sneakers, and radios. Desiring the Chicago Bulls and Teenage Mutant Ninja Turtles, Q'eqchi' youth value foreign commodities carried through transnational networks. By purchasing homes, stores, boats, televisions, bicycles, and land, the Q'eqchi' themselves are becoming owners. The most dramatic impact on their ideas and practice of morality is this recent transference of ownership.

Nonetheless, there are plenty of Q'eqchi' people who still blame the morenos. They tell me that crime in the villages is due to the fact the Q'eqchi' take note of what the Garifuna do, learning and copying their neighbors. This is why they now rape, steal, and scare people. As something learned, crime is not something they have intrinsically within themselves, as they claim for the Garifuna. Therefore, while the myth and historically influenced image of Garifuna as criminals has not changed significantly, Q'eqchi' people are now applying this identity to the youth and the budding capitalists amongst themselves. Politics, economics, and religion continue to create mythic-historical structures that allow Q'eqchi' to innovate roles and understand crime. These structures, however, are not rigid; they react to and are transformed by contemporary relationships of power.

ETHNOGRAPHIC CRIME

A friend visited me during the last few months of my fieldwork. We were walking to my house one night when a machete-yielding man popped out from behind a house and demanded our money. We did not hand over our money. I began to scream. I yelled that I lived in Livingston and that we were walking home and that he would be sorry for picking on the wrong people! Friends who still lingered at the corner of the street heard my cries and ran to assist us. Even though the man took off before the folks on the corner had a chance to see his face, they knew who it was. The next morning, a trail of people came to my house to tell me that they had spoken to him and that he was very sorry, that he did not know who we were. He thought we were tourists and sent us his apologies. Some asked us if we wanted a hit put out on him.

There is an unspoken social rule that sanctions tourists as the marks for crime. Bringing in dollars, disease, Western clothing, and lust, tourists are viewed as not giving anything back to the community. Many people insinuate that tourists deserve criminal violation. By robbing them, the people of Livingston have an opportunity to make the give-and-take bi-directional. Unfortunately, what happens by robbing the tourists is that the travel guides get filled with horror stories about Livingston's extreme crime rate and tourists stop coming, thus leaving a stagnant economy with no incoming source of revenue. Hotel and restaurant owners, who are predominantly Ladino or foreign, have formed an association to minimize crime and increase tourism. Many young Garifuna men, believing these hotel owners and outsiders are murdering known troublemakers in town, denounce the group as inherently racist—the association indiscriminately targets only Garifuna youth accused of criminal acts. Many people had told me that the young man who threatened my friend and me with the machete was on the top of the association's hit list because not only did he try to rob us and tourists but he also victimized locals to supply his crack habit.

We were white, so he saw us as an easy and accepted mark. What he did not know was that we were more than tourists. As an ethnographer, I had become a peripheral member of the community. I was economically and morally bound to the community. Even though I could pack my bags and fly through the sky to my foreign home, I had obligations to the people of Livingston. As an ethnographer, I froze lives through photography, excised pieces of culture from processes, and stole words and images for my own alien intentions. In that process, I also established relationships of trust and networks of exchange that were beneficial to the community. I was worth protecting.

Crime has been a constant in Q'eqchi' history. From the Spanish stealing treasures, Q'eqs robbing eggs, morenos lifting watches, to guerrillas demanding money, politicians breaking promises, and ethnographers reifying people, Q'eqchi' have continued to fit new circumstances into deeply embedded social and political categories. This, though, is not simply a color-coded racism or categorically driven process; this labeling of crime and criminals has to do with acts of morality—with what is and is not practiced. Criminals are defined by what they do, where they originate, who reaps the fruits of their labor, and with whom they collaborate or corroborate. Created through networks that stretch through time, across oceans, and through individual minds and collective histories, crime is naturalized and reinforced through extrapersonal structures, internalized frameworks, cultural practices, mythical and historical narratives, and everyday social exchanges. Employing their mythical, historical, and social relationships as a filter of perception, Q'eqchi' people identify, justify, and explain the transgressors who fly through their imagined world laden with foreign goods on their way to elsewhere.

NOTES

1. See Gonzalez (1988) for a complete history of the arrival of the Garifuna in Livingston.

2. Ferrapazco is the local name for *Empresa Ferrocarril Verapaz y Agencias del Norte, Ltda.*

3. Being a color-coded term, *moreno* is the Spanish term used in Livingston for the blacks.

4. *Reduccion* was a Spanish colonial policy of gathering indigenous people into organized town centers.

5. *Mandamiento* was a Guatemalan labor policy that forced landless Indians to provide work for the government and/or large foreign-owned agricultural estates.

6. See Cambranes (1985), King (1974), McCreery (1990, 1993, 1995), Sapper (1936), Wagner (1987, 1996), Wilk (1991), and Woodward (1990) for more history on the Q'eqchi' and the coffee industry in Alta Verapaz.

7. The Devil is also represented as having backward knees, legs, and elbows, thus demonstrating the Catholic influence on contemporary manifestations of Q'eq. Its Latin American and Caribbean manifestations, like the duende, are associated specifically with the history of conquest and colonialism.

8. See Comaroff (1996) for a cogent and concise discussion on identities as relationships.

9. Before arriving in Livingston, most Q'eqchi' people had seen the Q'eq in dances such as in *Baile de Katarina*. However, this dance has only one Q'eq, thus continuing his image as a solitary figure.

10. See the Comaroffs (1992) and Sahlins (1985) for the use of historical structures in understanding contemporary culture. Also see Holland et al. (1998), Strauss and Quinn (1997), and Ortner (1989) for more on how these structures become internalized filters of perception.

11. On December 29, 1996, the Guatemalan National Revolutionary Unity (URNG) and the government of Guatemala signed the Peace Accords, which finally put an end to the 36-year civil war.

BIBLIOGRAPHY

Bricker, Victoria
1981 The Indian Christ, The Indian King: The Historical Substrate of Maya Myth and Ritual. Austin: University of Texas Press.

Cabarrús, Carlos Rafael
1979 La Cosmovision K'ekchi' en Proceso de Cambio. San Salvador: Universidad Centramericana.

Cambranes, J. C.
1985 Coffee and Peasants: The Origins of the Modern Plantation Economy in Guatemala, 1853–1897. Woodstock, VT: CIRMA, Plumsock Mesoamerican Studies.

Comaroff, John and Jean
1992 Ethnography and the Historical Imagination. Boulder: Westview Press.

Comaroff, John
1996 Ethnicity, Nationalism, and the Politics of Difference in the Age of Revolution. In The Politics of Difference: Ethnic Premises in a World of Power. Edwin N. Wilmsen and Patrick McAllister, eds. Pp. 162–183. Chicago: University of Chicago Press.

Gonzalez, Nancie L.
1988 Sojourners of the Caribbean: Ethnogenesis and Ethnohistory of the Garifuna. Urbana: University of Illinois Press.

Holland, Dorothy, William Lachicotte, Jr., Debra Skinner, and Carole Cain
1998 Identity and Agency in Cultural Worlds. Cambridge and London: Harvard University Press.

Jenkins, Carol L.
1998 Ritual and Resource Flow: The Garifuna Dugu. In Blackness in Latin America and the Caribbean, Vol. 1, Norman E. Whitten, Jr. and Arlene Torres, eds. Pp. 149–167. Bloomington: Indiana University Press.

Kearney, Michael
1996 Reconceptualizing the Peasantry: Anthropology in Global Perspective. Boulder: Westview Press.

King, Arden R.
1974 Coban and the Verapaz: History and the Cultural Process in Northern Guatemala. New Orleans: Middle American Research Institute of Tulane University.

McCreery, David
1990 State Power, Indigenous Communities, and Land in Nineteenth-Century Guatemala. In Guatemalan Indians and the State: 1540–1988. C. Smith, ed. Pp. 96–136. Austin: University of Texas Press.

1993 Hegemony and Repression in Rural Guatemala, 1871–1940. In Plantation Workers: Resistance and Accommodation. B. V. Lal, D. Munro, and E. D. Beechert, eds. Pp. 17–39. University of Hawaii Press.

1995 Wage Labor, Free Labor, and Vagrancy Laws: The Transition to Capitalism in Guatemala, 1920–1945. In Coffee, Society, and Power in Latin America.

William Roseberry, Lowell Gudmundson, and Mario Samper Kutschbach, eds. Pp. 206–231. Baltimore: The Johns Hopkins University Press.

Ortner, Sherry B.
1989   High Religion: A Cultural and Political History of Sherpa Buddhism. Princeton: Princeton University Press.

Sahlins, Marshall
1985   Islands of History. Chicago: University of Chicago Press.

Sapper, Karl
1985 [1936]   The Verapaz in the Sixteenth and Seventeenth Centuries: A Contribution to the Historical Geography and Ethnography of Northeastern Guatemala. Los Angeles: Institute of Archaeology, University of California.

Stoll, David
1993   Between Two Armies in the Ixil Towns of Guatemala. New York: Columbia University Press.

Strauss, Claudia and Naomi Quinn
1997   A Cognitive Theory of Cultural Meaning. Cambridge: Cambridge University Press.

Taube, Karl A.
1992   The Major Gods of Ancient Yucatan. Washington, D.C.: Dumbarton Oaks Research Library.

Wagner, Regina
1987   Actividades Empresariales de los alemanes en Guatemala: 1850–1920. Mesoamérica 13:87–123.
1996   Los Alemanes en Guatemala: 1828–1944. Guatemala: Afanes, S.A.

Wilk, Richard
1991   Household Ecology: Economic Change and Domestic Life Among the Kekchi Maya in Belize. Tucson: The University of Arizona Press.

Woodward, Ralph Lee
1990   Changes in the Nineteenth-Century Guatemalan State and Its Indian Policies. *In* Guatemalan Indians and the State: 1540–1988. Carol Smith, ed. Pp. 52–71. Austin: University of Texas Press.

CHAPTER 2

CRIME AS A CATEGORY— DOMESTIC AND GLOBALIZED[1]

Laura Nader

It is a crime to kill a neighbor, an act of heroism to kill an enemy, but who is an enemy and who is a neighbor is purely a matter of social definition.
—E. R. Leach (1968:27)

INTRODUCTION

IN A SUCCINCT ARTICLE TITLED "Toward a Radical Criminology," published in the first edition of *The Politics of Law—A Progressive Critique,* William Chambliss (1982) juxtaposed the traditional question that criminology asked, "Why is it that some people commit crime while others do not?" with the sociology of law question, "Why are some acts defined by law as criminal while others are not?" (ibid.:230). He explains the reappraisal of the leading question as due to the 1960s civil rights demonstrations, anti - Vietnam War protests, and the media reporting on blatant criminality by political leaders and giant corporations. These, among other happenings, forced a reappraisal of criminology's focus on the individual and caused what Chambliss called a "paradigm revolution," encompassing the more broadly liberal understandings of criminal justice of the 1950s and 1960s.

In his paper Chambliss notes that what is criminal changes over time and that the political and economic forces behind the creation of criminal law is revealed in history. He gives some examples:

> Vagrancy laws reflected the tensions in precapitalist England among feu-
> dal landlords, peasants, and the emergent capitalist class in the cities . . .
> rights of rural village dwellers to hunt, fish, and gather wood were re-
> tracted and such activities became acts of criminality punishable by death
> as a result of the state's intervention on the side of the landed gentry in op-
> position to the customs, values, and interests of the majority of the rural
> population. . . . (ibid.:233)

At the same time, Chambliss is careful to take exception to the conflict theorists who argue that law is simply a result of ruling class activity. To fortify
his argument, he cites factory health and safety legislation that criminalizes
a business owner's refusal to comply.

Only a few years later, the second edition of *The Politics of Law* leaves
out Chambliss's dramatic rethinking of criminology and includes a new
piece by Elliott Currie called "Crime, Justice, and the Social Environment."
Moving away from the constructivist approach that Chambliss chronicled
for criminology, Currie speaks about the emergence of a conservative revolution in criminology, a revolution that defined crime as largely a criminal
justice problem. He dates the beginning of this conservative revolution as
appearing roughly from the beginning of the 1970s. Crime in the conservative view was caused by insufficient deterrence, that is, insufficient punishment. The major result of this conservative revolution, according to Currie,
is the rapid rise of incarceration in the United States (and the privatization
of new and old prisons, I might add).

Currie begins his 1990 version with the statement, "No one should
doubt that violent crime constitutes an American epidemic. Crime . . . (has)
brought tragedy and devastation to American cities, especially to poorer and
minority communities" (Currie 1990: 294). Indeed, the Sentencing Project
in Washington, D.C., has taken a look at blacks and crime. Their findings
are startling. One in three black men in their twenties are behind bars or
elsewhere in the justice system (Gest 1995). By some calculations the United
States imprisons black males at four times the rate of South Africa.

In his piece for the 1998 version of *The Politics of Law,* "Crime and
Punishment in the United States: Myths, Realities and Possibilities," Currie's
introductory sentence is even more dramatic: "The United States imprisons
its population at a rate from five to fifteen times that of other advanced industrial societies, yet we continue to suffer the worst levels of serious violent
crime in the developed world" (ibid.:381). He continues: " . . . more people
are murdered in the city of Los Angeles with 3.5 million people than in all
of England and Wales, with 50 million."

Currie's critique captures only part of the picture and returns to an earlier position that portrays crime as a social not a cultural problem. In this article he is not concerned with crime as a category. Little is said about crime in the corporate arena—the powerful destabilizing forces of rampant corporate crime—or the fact that the same forces behind prison privatization are now driving criminal justice policy, in one case lobbying that two strikes ought to result in lifetime imprisonment, rather than three, which would further inflate the numbers of people imprisoned.[2] In both the 1990 and 1998 versions of Currie's articles the emphasis is on reducing inequality, lessening poverty, developing family-friendly public policy, and so forth. So much for paradigm revolutions. It appears that we are now back to business as usual.

While the vast majority of public discourse in industrialized nations has not adopted a constructivist approach to crime, there has at the least continued to be a committed, if practically subdued, stream of publication and research on crime as a category. Furthermore, studies of the media's impact on perceptions and sentiments on crime have become increasingly available. Gary Potter and Victor Kapperler's edited volume *Constructing Crime: Perspectives in Making News and Social Problems* includes a range of essays exploring the flexibility of the nature and implications of "crime" in our increasingly media inundated society.[3] The application of theories and developments in the field of Critical Legal Studies has also found its way to criminology, with a number of journals, associations, and scholars devoted to the study.[4] But what impact has such work had on public perceptions?

At the outset, I should note that this article is not meant to be novel, although its configuration might be. Rather, it is intended to remind us of what we already know, because what we already know is critical to any broad scope understanding of crime in the United States today. Let me delineate the key points. First, the issue of boundaries and power is a key point in any assessment of academic or public policy considerations, one that should not wax and wane with political currents. That crime is a category applied arbitrarily in relation to social configurations expressed in law is illustrated by cross-cultural examination, and long ago accepted as an important finding among anthropologists. As my colleague Nelson Graburn used to say of the Canadian north, "No law, no crime." Or, as I will illustrate, environmental laws create new crime categories.

The second point relates to category boundaries and the observation that categories change. Where there is serious threat to the public good, as with hazardous work places and health and safety more generally, the charge may be civil or criminal. In spite of the near hysteria on street crime in the United States, new cases may indicate a shift in recognition of the seriousness of corporate and white collar crime, an awareness that extends the popular definition of crime beyond the "street," thereby raising comparable arguments on deterrence.

AN ANTHROPOLOGICAL FRAME

In 1990 I wrote an article on "The Origin of Order and the Dynamics of Justice." In it I noted that throughout American history corrective justice discourse, which values punishment, has often been found in opposition to distributive justice discourse, in which all individuals get their fair share. Proponents of corrective justice call for more stringent laws, stronger enforcement, and harsher punishments to prevent disruption of the social order, while their opponents champion social justice causes as the preferred means for creating order. Together these two views have dominated popular thought on crime, promoting an oscillation between government programs to cure law and order problems on the one hand and government programs to address questions of social injustice on the other. There has been little room for alternative thinking. I argued for a conceptual brake that would disturb this oscillation, and maintained that solutions grounded in order without justice, or social justice without profound social transformations, were destined to fail. Both of these ways of thinking about social order have proceeded without examining the ways in which, for instance, the misbehavior of corporations are excluded from discussions of crime. I argued that both the law and order people and the law and justice people use language that is supportive of the status quo, and that to address issues of "crime" requires returning to first principles, much as Chambliss attempted to do in his paper "Toward a Radical Criminology" (1982).

Toward this end, I return in this present work to the anthropological frame that emanates from ethnographic research in numerous world sites over time. The issues that have troubled sociologists and criminologists have also worried anthropologists, although the direction of anthropological thought has been more steady over time. Repeatedly, and no matter where anthropologists have worked in recent times, the question of native categories forces us to address the two powerful categories of Western law—"civil" and "criminal"—that are ipso facto part of our cultural baggage when we go elsewhere to work. The development of these categories in European cultures is related to the rise of the nation state and the need for the state to justify control and power over its citizens. But when anthropologists work in non-Western contexts we cannot simply accept the categories civil and criminal as given. In developing nation states they are clearly cultural constructs, the legacy of a specific Western tradition.

Presented with a society lacking in central authority, codes, courts, and constables, Bronislaw Malinowski's definition of *crime* (1926) was "the law broken," while A. R. Radcliffe-Brown (1933), a contemporary of Malinowski who had been influenced by Roscoe Pound's sociological jurisprudence, did not find terms such as civil law and criminal law useful. For all their differences, both Malinowski and Radcliffe-Brown viewed crimes as acts that cause the entire community to react in anger engendering a collective feeling of moral indignation.

In an encyclopedia article on comparative criminal law and enforcement (Nader and Parnell 1983), we concluded our assay of anthropological works with the same point: "The record on world societies has well illustrated that crime is a cultural construct" (ibid.:207), and the E. R. Leach quote at the start of this piece indicates widespread anthropological agreement. Although crimes, from the Western perspective, are violations of the law, violations of the law from the cross-cultural perspective are not necessarily crimes. The concept of crime, an idea related to Western jurisprudential history, becomes problematic when applied cross-culturally in societies with little or no "government."

In more recent work, most anthropologists do not define crime or attempt to use the distinctions between crime, tort, delict, sin, and immorality. That is, they eschew attempts to define crime in a universal manner. Their approach is particularistic, since categories in some societies may bear no resemblance to standard Western ones. On the other hand, it is also true that today all anthropologists work within state societies. Therefore, it is very likely that some behaviors anthropologists study will have been defined in criminal law as acts against the state. At the same time, anthropologists realize that state law and policy may not reflect the categories of thought of a socially differentiated population of citizens. In other words, political encapsulation brings into contact different systems of right and wrong and different ideas about how to treat wrongdoers. Present day examples in Indonesia and in Papua New Guinea (PNG) illustrate this point.

In an elegant description, David Hyndman (1994) analyzes how a state, faced with a debt crisis, favors investors who plunder natural resources and cast indigenous peoples in the role of subversive criminals. Anthropologists may view such "criminals" in a different light, as people taking up arms to protect their cultural and ancestral homelands, and to resist capital development. The Indonesian state and PNG, in collusion with transnationals, entered New Guinea to mine gold and copper. Hyndman calls this economic development by invasion. The cost of resisting invasion was heavy. Local peoples fought the foreign presence by blockading airstrips and blowing up pipes running from the mines; lives were lost and property was destroyed. Forced resettlement of indigenous populations often followed; local people became trespassers in their own land. Hyndman documents one invasion after another and notes ironically that Third World colonialism has replaced First World colonialism. Those people who resist notions of private property linked to capitalist development are considered criminals. Similar reasons for resistance arise in the fight for health and safety in the United States.

A CIVIL ACTION—
CATEGORIZING CRIMINOGENIC ACTS

criminogenic /'kri-me-no-je-nik "producing or leading to crime or criminality."
—*Webster's Third New International Dictionary.*
1986. Merriam-Webster, Inc.

Traditional tort cases include confrontations such as assault and battery or multi-vehicle collisions caused by intentional or negligent actions. Mass tort cases usually include a large number of victims, possibly even millions of claimants, such as in the Agent Orange herbicide case.[5] Mass toxic tort law is a relatively recent development in American law and is associated with growth in the industrial production of chemical substances in the last half of the twentieth century (Bloomquist 1996). Asbestos litigations and hormones such as diethylstilbestrol are examples of early toxic tort cases, each of which became celebrated in its own way. More recently, a nine-year legal saga that began in 1982 in Woburn, Massachusetts, has caught the attention of the public. The case focused on the period from 1960 through 1982, when children in Woburn, Massachusetts, exhibited an increase in adverse health effects.

Jonathan Harr (1996) is the writer who chronicled the Woburn case in the context of the workings of toxic tort litigation. His book, *A Civil Action,* is the story of how a group of families in Woburn became aware of a serious public health problem and what they wanted to do about it in relation to law. The story starts with one family whose three-year-old was diagnosed with leukemia and taken to Boston Children's Hospital and chronicles how over a short period of time the mother of this child realized that four children from her neighborhood were also diagnosed with leukemia. A newspaper article about the contaminated state of new town wells encouraged the mother and a local clergyman to organize a community meeting where it was learned that there were at least 12 cases of leukemia in Woburn. Such a clustering of cases was called to the attention of the Center for Disease Control and Prevention. Thus began a search for the cause of the leukemia. The families suspected that the Woburn plants of two corporations, W. R. Grace and Company of New York and Beatrice Foods Company of Chicago, had contaminated two municipal wells. Five of the families hired a Boston personal injury attorney; and, as the story unfolds, the reader begins to appreciate how difficult it is for plaintiffs to try to recover damages in a toxic tort case brought against corporate interests. The plaintiffs had the burden of proving that the contaminants in the well (specifically trichlocroethylene, or TCE) caused the children's leukemia, a connection that had not yet been scientifically proven.

An issue that remains to be addressed at another time is the full history of burden of proof; when it is shared by defendants and plaintiffs, when it is wholly the responsibility of the defendant or the plaintiff, and why, especially in relation to the Daubert principle[6] that is supposed to ensure adequacy of the scientific evidence. All of these issues kneaded together comprise a critique of *why there is so little tort litigation.* Not only does the plaintiff have uphill legal battles as illustrated by numerous cases, but massive campaigns of misinformation contribute to general ignorance about the American tort system—and where it stands in worldwide comparisons.

Harr's description of this extended Woburn case centered on the lead lawyer who enlisted numbers of experts in cancer epidemiology, hydrogeology, toxicology, geology, neurological studies, and more. The basic scientific research was being carried out while the case was being tried. The story highlights issues of class, culture, power disparities, awakened communities, and the place of a lawyer's perseverance and performance in a case against companies like Grace and Beatrice, powerful entitites eager to limit notions of corporate accountability. What originally looked like a medical problem became a public health problem and then a problem of the law. Complexity also became a problem in this case, as it is in other mass tort trials.

The trial was segmented—part of the managerial judge's move for economy and time-saving results. The experts refer to the separation of the interwoven issues of a legal case as polyfurcation. Some argue that polyfurcation of trials in complex tort cases could infringe on the Seventh Amendment, a plaintiff's right to a jury trial: "juries are forced by judicial and legal boundaries to hear only one part of the controversy and their ability to weigh links between the legal elements disappears" (Smith 1998). In the Woburn case, the judge trifurcated the trial such that the link between Grace, Beatrice, and the water, as well as the link between the water and the injuries, were presented separately, making it difficult for the jury to comprehend the story as a whole. The procedural rule permitting separation of issues clearly has impacts in relation to the outcome. In the Woburn case, keeping contextual information from the jury made it difficult for jurors to construct a valid picture of the entire case, a result that some legal experts argue "forces rational decision-making in juries while avoiding sympathetic or prejudicial decision-making" (ibid.:34). Sandra Smith quite correctly calls this "paternalistic treatment of jurors [which] leads to increased separation of issues" (ibid.). The assumption that jurors are not informed by the full context (criminal or civil) and that full context impedes rational decision-making is a subterfuge of the efficiency experts. There is no rationality in the absence of interrelatedness—rather, as in the Woburn case, confusion reigned among jurors who were expected to render a just verdict in a civil action.

Particularly because corporate interests have increased their attacks on the American tort system in the 1990s, Harr's story has current relevance. Contrary to public assertions about greedy litigants, this case makes one realize how difficult it is for aggrieved parties to win a case, even when litigants incur personal costs that could not have been higher. The costs of proving corporate wrongdoing are enormous. Besides, in this case, the plaintiffs most of all wanted an acknowledgment by the companies of the wrong they had done, and a personal apology to the families by the head CEO of one of the companies.

Ultimately, Beatrice won and W. R. Grace settled for eight million dollars. Each family received initially $375,000 and another $80,000 after five years, and there were expenses and legal fees. After the case was settled, the Environmental Protection Agency (EPA) issued a report citing Grace and Beatrice as responsible for contaminating the aquifer that fed the Woburn wells, and later the Woburn well field was slated for cleanup and placed on EPA's Superfund list. EPA also filed suit against Grace and Beatrice to collect the cost of the cleanup. But the bottom line in this case was that families filed a suit, a civil suit, against Grace and Beatrice because their children had been poisoned by the chemicals that had gotten into their water supply. Why was this a civil complaint? It could have been criminal—an offense against the state. But in the Woburn case the State of Massachusetts did next to nothing beyond closing the wells that were the source of the contaminated water. The EPA charged one of the companies, but the State of Massachusetts prosecuted no one.

MOVING CATEGORIES—A SOCIAL DEFINITION

The Woburn case could have been criminally tried, as was proven in two other recent Massachusetts cases (Alexander 1998). In *Massachusetts v. Fiengold*, a metals company and its chief executive officer entered guilty pleas in

November 1997 to charges of assault and battery with a dangerous weapon. Consolidated Smelting and Refining Co. and its chief executive were indicted in the Massachusetts Superior Court in Worchester County (BNA *Daily Law Report,* Nov. 25, 1997, A-8). They were charged with the criminal act of exposing company employees to lead dust and other hazardous chemicals. State environmental inspectors found that surfaces inside the company facility were covered with lead dust, and federal inspectors found the concentrations of airborne lead more than 200 times the United States government limit of permissible exposure. The company and the executive officer pled guilty to, among other charges, violating the state's Clean Air Act and illegally transporting hazardous waste. This case was the first of its kind in Massachusetts, but it sent the message to workers and employers that the boundaries between the civil and the criminal category were changing, and that pursuing criminal prosecutions against corporations could reduce the likelihood of workplace deaths. The stigma that accompanies a criminal prosecution is a more powerful deterrent than a monetary judgment that may be deducted from taxable income.

In a second case, *Massachusetts v. Hersh,* yet another metals company charged with assault and battery was accused of exposing workers to waste oil and three chemical solvents, two of which had been cited previously in the Woburn case. In this case, both the company and the executive officer pleaded guilty to disposing of hazardous waste in a manner that could endanger human health. Ultimately, the executive officer was sentenced to one year of probation and ordered to perform community service. There have been a dozen or so such cases from Massachusetts and elsewhere (Mokhiber 1999). In "Operation-Ill-Wind," 46 executives and 13 major defense corporations were convicted of defense procurement fraud. In addition, during the latter half of the 1990s, major firms such as Exxon, International Paper, Texaco, Nabisco, Weyerhaeuser, and Ralston-Purina have all been convicted of environmental crimes. Most recently, investigation of the tobacco and gun industries has illustrated how law can help expose the roles of corporate negligence and profit seeking in the creation of public health hazards.

The *Corporate Crime Reporter* (October 13, 1997) reports a string of corporations and executives that have been prosecuted for workplace deaths in recent years. The examples start in 1977 with the case of a Massachusetts fireworks company convicted of killing three workers after an explosion in an overloaded warehouse, and includes the case of Morton International and two of its supervisors who were to stand trial on charges of manslaughter in connection with the 1994 death of a worker who fell through a 60-ton pile of salt and was buried alive. An earlier example from 1992 involves the owner of a North Carolina chicken processing facility who pled guilty to charges of involuntary manslaughter and was sentenced to 20 years in prison for locking the escape exits from a chicken house that caught fire, resulting in the deaths of 25 people.

Susan Alexander (1998), a Chicago lawyer, compares these cases to the Woburn case and asks why the difference in what she sees as a quantum leap in the frequency of state and federal criminal cases against polluters, particularly in relation to hazardous workplaces. She cites the accumulation of new state and federal legislation regulating hazardous materials that began in the 1970s. She also notes that prosecutors, in addition to charging polluters with violations of environmental laws, are increasingly charging corporate defendants with assault and battery with a dangerous weapon. Alexander points out that such offenses have their origin in the common law but persists in asking, "Why didn't prosecutors file charges like this decades ago?" In response she cites complaints by workers, who are often the first to complain about hazards at the workplace. Alexander reports a study of 100 women workers funded by the Heinz Foundation in which one in four reported feeling that her health was endangered at work, while 28 percent were concerned about hazardous materials. She is correct in noting that legislation protecting whistle blowers has probably encouraged workers to voice their complaints. The media has focused attention on hazardous pollution, and the public is increasingly outraged by corporate wrongdoing in relation to contamination of the air and water.

What Alexander speaks about may be a quantum leap in the United States, but in societies where communities are in closer touch with issues of survival, the notion that people who endanger the public health of communities should be held responsible is much more widespread. In such places, the idea of public accountability is equally strong. Thus, assertions that the pendulum has swung too far in the direction of protecting the environment appears ludicrous in comparative perspective. One example follows.

SMALL DEMOCRACIES

El respeto al derecho ajeno es la paz (Peace lies in respecting the rights of others).
—Benito Juarez

It has always interested me to read northern discourses about the need to export democracy to the rural areas of Latin America because small face-to-face communities, especially indigenous communities, often practice a good deal of democracy. In my study of local government and the practice of law among the mountain Zapotec (Nader 1990b), I chronicled many of these village democratic practices. I remember that early in my fieldwork I was curious about what these mountain Zapotec thought was a serious legal case as compared to something less serious. For example, it struck me as counter-intuitive that theft should be considered as serious as killing someone. Yet, for the Zapotec, it was, particularly if the thief was stealing food harvest.

At the time, my investigation proceeded along two lines—the collection and comparison of cases and what was done about them, and interviews with the judges (e.g., the *presidente*) or decision makers. It might be useful for me to reiterate some of what was revealed by these modes of exploration. The first interview that I conducted with an experienced presidente proceeded in the usual question and answer form. "Can you tell me of a serious case you had?" The presidente then began to tell me about the man who robbed a water drum (Nader 1990b:99). When the man was formally accused of steal-ing a water drum, he denied the charge. When confronted with the evidence, however, he had to pay a stiff fine, and soon thereafter he left the village and moved to the state capital. The man accused of theft was a well-to-do fellow who did not *need* to steal for money, as the presidente said. He stole, the peo-ple thought, for ambition; and, after being jailed for one night and after pay-ing his fine he could not face his neighbors, and so he left.

When I asked if there were other serious cases besides theft, the presidente told about the altercation he had with the village police who wanted him to levy heavy fines against those committing crimes in the streets. As the presidente said: "I saw that the offenses were not very serious, so I paid little attention to them (the police). I fined the defendants according to what they had done." And what had they done I asked? "They had fought, yelled, drank beer . . . or did nothing more than make noise." The police wanted the presidente to put them in jail and fine them heavily, but the presidente refused. The next day the police took the case to the higher level state district court; their complaint was against the presidente and they asked the judge to give them guarantees. It was a dispute over who had authority—the police or the presidente. The presidente explained, "They wanted me to jail these men because they were drinking beer. They wanted me to follow them, to show me, but I did not want to do so. . . . They became angry" (ibid.:101). Finally, in answer to the question about why theft was so serious, I asked: "Is it more serious than killing?" His response: "About the same, because if you would steal millions of pesos, it is like killing these people. You leave them without clothes" (ibid.:104).

In one of the first pieces I published on the Zapotec (Nader 1964), I used the range of sanctions as an index of seriousness of the charge. Low-ranking sanctions were applied to offenses such as street fighting, drunkenness, dis-turbing the peace, nuisance to the court, failure to comply with municipal obligations, and other such matters. More serious sanctions were meted out to those accused of abandonment, abduction, assault and battery, attempted murder, slander, theft, and boundary trespass. One might also include among the more serious complaints abuse of authority, contamination of drinking water, and problems over public use of public spaces. For these Zapotec people, endangering the interests of the Commons is among the most serious cases that are referred to the court.

What would the Zapotec say about Woburn? Most likely they would debate among themselves the charge that two groups contaminated the drinking water and that this resulted in eight poisoned children who died. All sorts of facts would be admissible because context is critical to decision making. It would probably not be difficult to find out whether the two accused groups were guilty because, in a small community, it is hard to hide one's business. The case would likely be about murder. But, despite their sophisticated legal reasoning, the Zapotec would be in the same place or worse than the families of Woburn should a modern case of pesticide poisoning actually occur. First, the case could not be heard in the village; it would be taken to the district court or Oaxaca City, the state capital. Secondly, the accused companies might not be national companies; they could be international or multinational and beyond the law. Finally, small agricultural communities would not have the resources to pursue such litigation even if there were laws protecting them and a Mexican version of the EPA. The story of such an attempt to bring multinational corporations to justice would be brief, which gives us some sense of the incentive needed to get to where we in the United States are in tort litigation cases.

If one wants to chart the evolution of tort litigation, and especially mass toxic tort cases, it would be a story of the elongation of the distribution chain in which latency and the time dimension plays a large role—both the time that it takes for harm to surface and the time that it takes for scientific evidence to catch up with the uncovering of public health harms. As depressing as the story of Woburn is to many American readers, the United States is probably the leading industrial country in the development of tort litigation. Yet the challenge here as in any Western legal democracy returns to the challenge of crime as a category. Is a criminal charge more effective than a civil suit? Is it easier to prosecute? And what about differences in burden of proof between civil and criminal cases?

MODERN LAW AND THE QUEST FOR RESPONSIBILITY: TWO IRONIES

There are two ironies relevant to the subject of crime as a category. The first involves the stigmatization of the poor and powerless: the second pertains to the incivility involved in the pursuance of a civil complaint. In the first instance, we are bombarded with statistics about street crime. Yearly, the Federal

Bureau of Investigation (FBI) publishes its Uniform Report for Crime in the United States. The report documents murder, robbery, assault, burglary, and other street crimes, while ignoring corporate and white-collar crimes such as occupational homicide and life-harming industrial contaminations. Russell Mokhiber (1999), editor of the *Corporate Crime Reporter*, points to evidence that indicates that crimes committed by people of means cost the nation about 50 times as much a year as the combined cost of burglary and robbery. He was quoting from a report by a professor of accountancy at Brigham Young University. He goes on to note that the FBI publishes a homicide rate for street crime that is only half as high as the numbers the Labor Department reports for occupational diseases. Mokhiber notes that "On-the-job homicides are some of the most heinous crimes corporations could be charged with. Yet corporate violence that results in worker deaths rarely provokes criminal prosecutions, either at the state or federal level." Mokhiber also quotes a number of Washington reporters who declare that young black males commit most of the crimes in Washington, D.C. It is because the stigmatization that accompanies the criminal category is so powerful that we have had such powerful lobbying against the criminal category for certain acts and for the "violation" category instead, something Edwin Sutherland wrote about many years ago (1949). From this flows another irony that pertains to the pursuance of justice.

It has been observed that those who commit corporate misdeeds are commonly those who, unlike all other criminal groups in the United States, have the power to define the law under which they live (Mokhiber 1999: 14). Mokhiber uses the federal auto safety law as a case in point, noting that the law carries no criminal sanctions. The auto industry lobby has for years blocked the passage of laws that would add criminal sanctions to corporate offenses, at the same time as they blocked the law requiring airbags in all new vehicles. Estimates are that "almost three Vietnam walls worth of Americans" died in auto crashes since the early 1970s because the auto companies thwarted all efforts to develop and legally mandate the air bag in American cars (until the air bag law was passed in 1991). Mokhiber also notes that even if there were criminal sanction laws, the problem of prosecution would be central because again, unlike most other criminal groups, powerful corporations can influence prosecutors not to bring criminal charges. The social science literature on prosecution is replete with references to unacknowledged class and racial biases, poorly drafted laws, political deals, and incompetence as factors that mitigate against the criminal prosecution of the powerful. Mokhiber calls the problem "law enforcement obscenity" and points out that someone who harasses an animal gets more time than someone who violates federal worker safety law. He explains that, "The maximum criminal penalty for harassing a wild burro on federal land is one year in jail, and seven people have been jailed for this crime" (1999: 15). Clearly, use of the criminal category, application of criminal sanctions, and, apparently, the public's general perception of the seriousness of an offense, all are mediated by the diacritics of power in the relationship between the perpetrator and the victim.

In *The Growth of American Law*, Willard Hurst (1950) sees law as a term of convenience without any precise boundaries, and his Wisconsin external legal history school takes seriously Oliver Wendell Holmes and Roscoe Pound's call for reconnecting legal to social history (Gordon 1975). Theirs is not an autonomous law, but one that sees law as connected to social structure, administration, economic and political organization, professional habit, or religion. However, the widespread notion of the law as distinctively legal, free from the effects of social stratification and political power, has had a pernicious effect. Perhaps only a nonfiction writer or a good ethnographer could reconnect the legal to the social and cultural as did Jonathan Harr—providing what in another context someone called democratic detail, rather than aristocratic detail. There is more to understanding mass toxic tort cases than the conceptual aspects of the adversary process. Other actions by the state, the politicians, the regulators, the propagandists, the harmed, and the harmers are all part of the law making process; so too is crime as a social category. The internal analysis of law stays within the box of adversariness. The anthropological perspective, along with Willard Hurst, concentrates on interactions between the box and the wider society to explore social context and the social effects of law. Thus, while the idea of "equality before the law" obscures the fact that there is rampant social inequality, equality before the law is a legal guarantee for the protectors of citizens from real power differentials that affect or might affect them. The Anglo common law tradition states that legal decisions should be based primarily on legal precedents, which places a tremendous social import to each individual case. Under this interpretation, law is constructed from below, and the legal classification that differentiates criminal from civil cases may be reinterpreted from the point of view of new interest groups.

CHEMICAL RIGHTS

Our administration of justice is not decadent. It is simply behind the times.
—Roscoe Pound (1906:403)

An increase in incidents that seriously threaten the public good accompany the growing problem of industrial chemical contamination. Anthropologists encounter such cases in the places where they have traditionally been working. For example, Canadian studies have shown that the Inuit people on Broughton Island, which is part of the Baffin Island group in Canada's Arctic Northwest Territories, have the highest levels of PCB found in any human population (Colborn 1996).[7] Other Baffin Island Inuit shun the villagers as the "PCB" people and refuse to intermarry, as if they themselves were clear of contamination. The victims of corporate irresponsibility, rather than the perpetrators, are banished and shunned. Industrial chemicals not only contaminate, they also change human biology and transform social and cultural forms.

Chemicals have merged with human qualities in other ways as well. Here in the United States, the law reflects anthropomorphic qualities chemicals have acquired. For instance, under United States law, chemicals have rights (Coco 1998). That chemical rights precede human rights is indicated in numerous Supreme Court cases dealing with constitutional law and industrial rights. A chemical is assumed harmless until proven harmful. Chemicals enjoy certain legal privileges and protections under the Toxic Substance Control Act enacted on October 11, 1976. The EPA charges the Office of Population Prevention and Toxics with the responsibility of proving that some chemicals present an "unreasonable risk" to human health or the environment. The cumbersome procedure for making such judgments, a process that entails comparing risks with supposed economic benefits, invites negotiation. According to one study of "chemical rights" (quoted in ibid. 1998:16): "The EPA representatives do admit that industry can negotiate the entire agreement. . . . The EPA comes to the table asking for things they know that they cannot get. . . . It's just like negotiating a settlement in a tort case."

The larger story is about toxic deception or how "the chemical industry hides behind the facade of science, misuses the law, and endangers your health" (Fagin and Lavelle 1999:14). Coco concludes that "Regulation of corporate behavior under the TSCA (Toxic Substance Control Act) has been for the most part turned on its head. The rules are working to protect chemical companies not to protect the public" (ibid.:38). She advocates overturning the chemical rights doctrine, and the use of public relations campaigns to do so. If industry wants to contain the debate within the language of risk assessment, create a user-friendly public database such as the 1998 EDF (Environmental Defense) launched *Chemical Score Card* information service on the Internet. At the moment, however, the general population serves as the laboratory for discovering adverse health and environmental effects.

CONSTRUCTING CRIMINALS WORLDWIDE: POSSIBILITIES FOR PARADIGM SHIFTS

For a brief moment in 2001, the shockwave of Enron's collapse on the financial well-being of its employees as well as the consciousness of American politics seemed to offer an aperture for change.[8] The dismay of the public at the gross, and to this day unrepentant, corporate abuses of what was once considered a model U.S. corporate player offered the opportunity to bring white-collar crime to the forefront of the discussion on law and order in our criminal justice system. But, for all of the outrage, considerably little has been done to effect structural changes in our system of corporate legal governance. Enron has been whitewashed as an epiphenomenal error, a particularistic occurrence that does not demonstrate any systemic flaws in how we regulate corporate crime.[9] Any challenge to the line between civil and criminal law was not fully sustained. While certain offenders will continue to be imprisoned under the three strikes law for astronomically less damaging crimes, Enron executives have not had to worry about the same fate.

Within the same period of time, debates have engaged the issue of the legal status of "terrorists." How one should define acts of terrorism under existing international law and the Geneva conventions was implicated in the imprisonment of Al Qaeda members captured in Afghanistan by the U.S. military after the 9/11 attacks in the United States.[10] While the Bush Administration argued that these persons were in fact beyond the protections of any law, Amnesty International claimed they should be treated as prisoners of war. The American Jonathan Walker Lindh (who was caught amidst the Taliban), further complicated these deliberations, as his crime would have been differently conceived depending on his status as an American citizen. This gets to the heart of Leach's quote at the outset of this paper about who decides where the line is drawn.[11] The debates were ultimately mute, for the power to finally prescribe the status of the detainees is in the hands of the Administration, regardless of the merits of the case or international deliberations.

However, while the domestic effects of the Enron and detainee cases are still being played out, what both instances have in common is that the discretion to categorize certain behaviors as criminal is no longer simply an issue of internal domestic jurisdiction. Enron is a multinational corporation whose fate was tied extensively to its investments in India. The attack on the World Trade Center occurred on American territory, but involved foreign citizens who represented an international organization. This raises issues beyond the traditional criminological conceptualization of space. It is only within the last decade that a substantial literature has arisen around the "globalization of law" that has complemented, if not predicated, the globalization of our economy.[12]

What this literature demonstrates is that in the arena of international interactions, significant cultural barriers exist in the field of legal ideas.[13] While a great deal of scholarship has been concerned with the interactions between internationalized business, the literature seems to point to an even greater disjuncture in ideas about criminality. Moreover, the power asymmetries that exist at this level of legal interaction limit the discursive and material possibilities for a consensus on justice-related issues. The recent withdrawal of the United States from the International Criminal Court treaty, as well as its promotion of alternative dispute resolution over adjudication in international contexts, is one indication of this dissonance.[14] The varied domestic application of the term "terrorist" by a range of nation-states also demonstrates the relativity of criminal categories in different social contexts.[15]

Ultimately, these types of cross-national legal conflicts will require more democratic thinking for a constructive dialogue in a globalized legal field where public law is often far less developed than private commercial law.[16] Nonetheless, engaging the difficulties that confront internationalized laws may eventually provide a path toward reassessing legal practice at the domestic level than is currently provided for in popular law and order debates. In other words, an adequate theory that explains why some acts are defined by law as criminal while others are not might serve to shift the current civil and the criminal paradigm toward consequence thinking rather than rigid adherence to categories.

NOTES

1. An earlier version of this paper was published in *2001 Windsor Yearbook of Access to Justice,* Vol XIX. My thanks to Jed Kroacke for his help in preparing this revision.

2. From the *California Criminal Law Observer:*

   In 1994 California voters approved a ballot initiative known as "Three Strikes and You're Out." Basically what it means is that people who are convicted of three felonies may end up facing life in prison. The actual "law" has five major moving parts. First there is the ballot initiative (i.e., Proposition 184), then there is there the actual statute that was passed [California Penal Code Section 667 (b) through (i)], and then there are three other code sections that identify the types of violations that count as "strikes" against you.

   From the American Bar Association, Division of Media Relations and Public Affairs:

   Facts About the American Criminal Justice System: What has been the impact of "three strikes" laws on crime and the criminal justice system?

The impact is unclear. The only real research comes from California, because that is the only state that is making frequent use of the law. While at least 22 states and the federal government have enacted three-strikes laws since 1993, the laws in most jurisdictions are drafted much more narrowly than in California, and for this reason, or because they have not seen the need, prosecutors nationwide have not extensively applied three-strikes legislation. How has "three strikes" legislation worked in California? The vast majority of those sentenced under the law—85 percent—are sentenced for nonviolent offenses. And second and third strike cases are resulting in many more jury trials. While more than 90 percent of felony cases are disposed of through plea bargaining, many fewer offenders agree to plead guilty in three-strike cases. Such cases account for only three percent of the filings in Los Angeles, but make up 24 percent of the jury trials. A California study also found that African Americans—who make up seven percent of the state's population and 20 percent of its felony arrests—are imprisoned under the law 13 times as often as whites, and constitute 43 percent of third strike inmates.

3. Also see Surett (1998).

4. Not unexpectedly, this type of scholarship has not had a warm reception in many academic articles. Bruce Arrigo (2001) details the barriers that critical criminology has faced and the specific responses of leading journals to submissions from this field.

5. From Tokar (1998):

   The herbicide "Agent Orange," which was used by U.S. military forces to defoliate the rainforest ecosystems of Vietnam during the 1960s was a mixture of 2,4,5-T and 2,4-D that was available from several sources, but Monsanto's Agent Orange had concentrations of dioxin many times higher than that produced by Dow Chemical, the defoliant's other leading manufacturer. This made Monsanto the key defendant in the lawsuit brought by Vietnam War veterans in the United States, who faced an array of debilitating symptoms attributable to Agent Orange exposure. When a $180 million settlement was reached in 1984 between seven chemical companies and the lawyers for the veterans, the judge ordered Monsanto to pay 45.5 percent of the total. In the 1980s, Monsanto undertook a series of studies designed to minimize its liability, not only in the Agent Orange suit, but in continuing instances of employee contamination at its West Virginia manufacturing plant.

6. In *Daubert v. Merrell Dow Pharmaceuticals,* plaintiffs sought to introduce evidence that birth defects had been caused by a mother's ingestion of the antinausea drug Bendectin. Despite the fact that the Supreme Court sided with

the plaintiffs in *Daubert,* by setting up a process for judges to determine the reliability and relevancy of scientific evidence, the decision has been perceived by many trial lawyers as giving trial judges important powers to exclude scientific evidence and thus undermine plaintiffs' chances in toxic tort cases.

7. For source of PCB in Baffin Island, see Colborn (1996).

8. Robin Sherril is a frequent contributor to *The Nation*'s coverage of corporate crime and related news stories. Along with Greider's (2002) "Crime in the Suites" (specifically dealing with Enron), Sherril's articles are good commentary that supplement the weekly summaries of the *Corporate Crime Reporter.*

9. Although often forced to be published by small presses, work has continued to raise the issue of corporate crime. See Mokhiber (1999) and Mokhiber and Weissman (2001).

10. The American Friends Service Committee has developed a resource detailing the various positions on the detainees with links to various international legal documents and statements by U.S. officials and various writers. This can be found at: http://www.afsc.org/nomore/detainee.htm.

11. In lectures delivered at the University of Edinburgh in 1977, E. R. Leach directly addressed the links between the rules of war and the categorization of certain crimes as terrorism.

12. See Dezalay and Sugarman (1995). This volume in part analyzes the roles that lawyers and law plays in the internationalization of capitalism.

13. In *Law, Capitalism, and Power in Asia,* Kanishka Jayasuriya (1999) and other contributors look at the imperfect reception of the "rule of law" in Asia. Jayasuriya argues that cultural specificities always rework the reception of legal institutions and that domestic legal cultures can conceive of the legal in fundamentally different ways. Also relevant is Findlay (2000).

14. See Nader (1997, 1998).

15. The Macedonian government lobbied unsuccessfully to categorize the Albanian guerrillas as "terrorists" in line with the U.S. "war on terrorism." The United States and NATO did not agree with this characterization. Merits of the classification aside, the disagreement involved a great deal of political capital in light of U.S. recent policies.

16. Dezalay and Garth (2002) give a detailed explanation of why developments in the legal delivery of public goods lag far behind harmonization in international business law, most specifically in Latin America.

BIBLIOGRAPHY

Alexander, S.
1998 Getting Tough in Fight Against Pollution. Chicago Daily Law Bulletin. March 11.

Arrigo, Bruce
2001   The Perils of Publishing and the Call to Action. <http://www.critcrim.org/
        critpapers/arrigo_pub.htm>
Bloomquist, R. F.
1996   Bottomless Pit: Toxic Trials, the American Legal Profession, and Popular Per-
        ceptions of the Law. Cornell Law Review 81: 953–988.
Chambliss, William
1982   Toward a Radical Criminology. In The Politics of Law—A Progressive Cri-
        tique. D. Kairys, ed. Pp. 230–241. New York: Pantheon Press.
Coco, L.
1998   Chemical Rights. University of Maryland Law School, unpublished manuscript.
Colborn, Theo
1996   Our Stolen Future: Are We Threatening our Fertility, Intelligence, and Sur-
        vival? New York: Dutton.
Currie, E.
1990   Crime, Justice, and the Social Environment. In The Politics of Law—A Pro-
        gressive Critique, D. Kairys, ed. Pp. 294–313. New York: Pantheon Press.
1998   Crime and Punishment in the United States: Myths, Realities and Possibili-
        ties. In The Politics of Law—A Progressive Critique, D. Kairys, ed. Pp.
        381–409. New York: Pantheon Press.
Dezalay, Yves and Bryant Garth
2002   The Internationalization of Palace Wars: Lawyers, Economists, and the Con-
        test to Transform Latin America. Chicago: The University of Chicago Press.
Dezalay, Yves and David Sugarman, eds.
1995   Professional Competition and Professional Power: Lawyers, Accountants and
        the Social Construction of Markets. New York: Routledge.
Fagin, Dan and Marianne Lavelle
1999   Toxic Deception: How the Chemical Industry Manipulates Science, Bends
        the Law, and Endangers your Health. Secaucus, NJ: Carol Publishing Group.
Findlay, Mark
2000   Globalization of Crime: Understanding Transnational Relationships in Con-
        text. Cambridge: Cambridge University Press.
Gest, T.
1995   A Shocking Look at Blacks and Crime. U.S. News and World Report
        119(15): 53–54.
Gordon, Robert W.
1975   Willard Hurst and the Common Law Tradition in American Legal Histori-
        ography. Law and Society 10(1): 9–55.
Greider, William
2002   Crime in the Suites. The Nation, February 4: 11–14
Harr, J.
1996   A Civil Action. New York: Random House.

Hurst, James Willard
1950 The Growth of American Law: The Law Makers. Boston: Little Brown.
Hyndman, David
1994 Ancestral Rain Forests and the Mountain of Gold—Indigenous Peoples and Mining in New Guinea. Boulder: Westview Press.
Jayasuriya, Kanishka, ed.
1999 Law, Capitalism, and Power in Asia. New York: Routledge.
Leach, Edmund R.
1968 Ignoble Savages (a review of four books on violence). New York Review of Books 11(6): 24–29.
1977 Custom, Law, and Terrorist Violence. Edinburgh: Edinburgh University Press.
Malinowski, Bronislaw
1926 Crime and Punishment in Savage Society. New York: Harcourt Brace. Reprint New York: Humanities Press. 1951.
Mokhiber, Russell
1994 Underworld, USA. In These Times, April 1: 14–16.
1999 Corporate Predators: The Hunt For Mega-Profits and the Attack on Democracy. Monroe, Maine: Common Courage Press.
Mokhiber, Russell and Robert Weissman
2001 On The Rampage: Corporate Power in the New Millennium. Monroe, Maine: Common Courage Press.
Nader, Laura
1964 An Analysis of Zapotec Law Cases. Ethnology 31(4): 409–419.
1990a The Origin of Order and the Dynamics of Justice. In New Directions in the Study of Justice, Law, and Social Control. John R. Hepburn, ed. Pp. 189–206. New York: Plenum.
1990b Harmony Ideology: Justice and Control in a Zapotec Mountain Village. Stanford: Stanford University Press.
1997 Civilization and its Negotiators. Current Anthropology 38(5): 711–37.
1998 The Globalization of Law: ADR as "Soft" Technology. Proceedings of the 93rd Annual Meeting, Pp. 1–9. Washington D.C.: American Society of International Law.
Nader, Laura and Philip Parnell.
1983 Comparative Criminal Law and Enforcement: Preliterate Societies. Encyclopedia of Crime and Justice (Volume 1). S. Kadish, ed. Pp. 200–207. New York: The Free Press.
Potter, Gary and Victor E. Kappeler.
1998 Constructing Crime: Perspectives in Making News and Social Problems. Prospect Heights, Ill.: Waveland Press.
Pound, Roscoe
1906 The Causes of Popular Dissatisfaction with the Administration of Justice. American Bar Association Report (Part 1) 29: 395–417.

Radcliffe-Brown, A. R.
1933 Primitive Law. *In* Encyclopedia of the Social Sciences (Volume 9). E. R. A. Seligman and A. Johnson et al., eds. Pp. 202–206. New York: Macmillan.
Smith, S. A.
1998 Comment: Polyfurcation and the Right to a Civil Jury Trial: Little Grace in the Woburn Case. Boston College Environmental Affairs Law Review 649.
Surett, Ray
1998 Media, Crime and Criminal Justice: Images and Realities. Pacific Grove, CA: Brooks/Cole Pub. Co.
Sutherland, Edwin
1949 White Collar Crime. New York: Dryden Press.
Tokar, Brian
1998 Monsanto: A Checkered History. The Ecologist, September/October: 254–280.

CHAPTER 3

THE ANTHROPOLOGIST ACCUSED

June Starr

Editors' Note: June Starr passed away before this volume went to press. This essay, one of the last she wrote, is a painful and dramatic analysis of the ways that ethnographers' personal lives are often put on the line in fieldwork. It is a testament to her courage and commitment. She was a dear friend and colleague to many contributors to this volume.

June wrote this chapter as she looked back on her initial ethnographic work in Turkey through the years of experience that followed. She therefore speaks in several voices, including that of a narrator—an anthropologist and professor of law, that of a woman contemplating her younger self, and the voice of a graduate student in anthropology facing the challenges of field research in turbulent times.

PREFACE[1]

The 1960s was the heyday of cultural anthropology in the United States as departments expanded across the nation and considerable governmental support existed for graduate student field research. To obtain a doctoral degree in anthropology, it was mandatory for a student to spend a year or more engaged in empirical field research in a culture not his own. This emersion in a foreign culture was supposed to transform the person from his former self into an anthropological person because, through field research in a foreign place, he made himself vulnerable to the native people. Some anthropologists suggested the fledgling anthropologist learned the culture like an infant of that place, slowly acquiring the cultural rules for behavior by making mistakes and being corrected by natives.

The most prestigious field sites were among tribal peoples living in a situation of stone age technology and economy along the Amazon or in the Philippines. They were remote villages in little studied parts of the world, where non-European languages were spoken. Many felt that "authentic culture" unaffected by modernity could be found in these places, giving the anthropologist an opportunity to study social systems that had evolved without the influence of Western cultures.

The classic Western monograph for Turkey, called Turkish Village, *had been written by the English anthropologist Paul Stirling (1965). A Turkish anthropologist, Ibrahim Yasa, had written a village study,* Hasanoğlan, *which had been published in English in 1957. And because Turkey had a flourishing intelligentsia, there were also a number of village studies in Turkish by Turkish social scientists. Starr set out to conduct the first study of the processes of settling disputes in rural Turkey. By choosing western Turkey as the place to find a village, Starr was locating her research in an area where no previous village studies had been conducted. Once in Turkey, she discovered one or two other social scientists had worked in western Turkey—but not enough to deter her from this region.*

It happened a long, long time ago. Long before laptop computers, cellular phones, and e-mail and fax. Long before the Concorde, car seats for toddlers, and windsurfing. Long before people made phone calls across the continents in the blink of an eye. If you wanted to "reach out and touch someone," he had to be standing in the same room with you.

We had not yet become a celebrity culture, although we were already mesmerized by Jackie Kennedy's glamour and poise, especially at the ceremonial funeral of her young husband, the president. We had not yet learned to mistrust our government through the Pentagon Papers or the Watergate tapes. We were still in the preglobal economy.

I have lived more passionately since then, and even more recklessly near the edge, but never so far away from my own culture that I could not find a bridge back. There, in that time and place, there were no bridges. There were almost no viable choices for the June I was then. If she wanted to survive psychically, the circumstances had to be *lived* out. The person I am today would not have done it as she did. She made major mistakes, first by not taking action, then by acting, but I am the kind of person who would never have gotten into such a quagmire. I am now a careful planner. I am much less of a romantic. I see people more clearly. I have learned not to count on them. I walk away from those who want to play mind games with me. I have lived in troubled cities, but have never lived in such seemingly beguiling circumstances as when the personal and the historical intersected in Turkey.

Call this a confession, if you must. I, of course, do not think it is. It is an objective assessment, a narrative, an account. We first see events (and even her, the younger me) from the outside, from the knowledge of knowing how her choices turned out. Then we see one very brief internal glimpse of her as she begins her anthropological field life and first meets Adnan. Finally, we read excerpts from her "Report of Events to the Anthropology Faculty, University of California at Berkeley in spring 1968." Here we find that the very actions that gave relief to the village people, her finding a Turkish male field assistant, Adnan, to help her with her work and difficult living situation, provoked the Turkish authorities. The authorities viewed Adnan as a more sinister figure than she did—as someone who enabled her to engage in illicit activities.

Will you understand her? Ought we to be judging her? She lived in such different times; she felt she had so few options. She felt that it had taken all her energy and psychic resources just to get this far, and she had to make it work. She had staked her identity as a person on her identity as an anthropologist. I should not really be judging her. I have published two books and a film from her fieldwork materials.

Yet, she tripped up in her first field research as a legal ethnographer. The way she "inhabited" the role of the "anthropologist in the field" was unpersuasive to the Turkish authorities. As her identity as *wife* slipped away, the Turkish authorities, never entirely subtle regarding affairs of the heart, found an international metaphor for her. She was a spy, living in a Turkish village. The charges were never made formally, but they were in the air, swirling around her as she did her work. Was she a spy? Of course not. She was a young woman in a Muslim culture on her first anthropological expedition whose husband had just left her in the rural countryside. As her field research progressed, could she be called a spy in the Turkish court system? Perhaps, but she only studied ordinary disputes. She knew nothing of political trials, Turkish politics, or ethnic animosities. Was she a spy into local Turkish preparedness for war against Greece? She tried not to be. When huge cannons were brought into the coastal villages and aimed at Greece, she began a legal survey of the mountain villages up north, far from the Aegean coastline. Was she an archaeologist digging for gold treasure in the ancient fields around Bodrum? No one saw her or anyone connected to her with a shovel in hand, except when her Land Rover sank into the sand at the seaside village and had to be dug out. What is known of her is that she staked her life on being accepted in her role as an anthropologist. And yet, later, much later, she lost her way. For this comes to you from the distance of thirty-some-odd years, a different discipline, and a different profession.

BEING AND NOTHINGNESS

Imagine a person, a woman, who upon waking each morning for four years thinks, "Is this the day I will leave him?" Imagine a young woman in a marriage that engulfs her yet seems unreal. Sometimes there were pleasant and easygoing times; she lived a pleasant and dull life. He made all the major decisions—where to study when they were graduate students, where to live after his Ph.D., what house to buy, where to take vacations. Thus, their marriage was characterized by what Betty Friedan has called that "thing with no name."

With great difficulty a child was born to this couple. Because of the child, the woman gradually began to imagine shaping a future away from the shadow world in which she lived. Gradually, she began to define normalcy as going to the university, attending graduate classes, writing papers, taking exams, and moving toward a graduate degree.

He had not wanted her to become a doctoral student; sometimes, he still became angry after dinner, scolding her, yelling at her, and leaving the house in the middle of the night, sometimes staying away for several days. She never got used to his anger nor understood its causes. But, in the end, she prevailed. She wrote research proposals, took exams, had coffee with other students, and made a tentative plan with her husband. They would go together to Turkey for her rural field research. He would use his year's sabbatical to work on his second book. She would study how Turkish villagers resolved disputes and quarrels among themselves. They would travel a little together. They would have a new beginning, not unlike one of the several new beginnings they had undertaken since their marriage started.

He began attending Turkish language classes. Having a good ear and a nimble tongue, he quickly became fluent. Her plan was to learn Turkish through someone who knew a little English. She had taken linguistics courses on how to learn a foreign language, using English as a contact language. Also, she had acquired Turkish language books and passed a reading exam in Turkish.

Thus, several weeks before Christmas in 1966, a year after their joint decision, they found themselves in western Turkey in a red Land Rover, at the unpaved and muddy Canakkale ferry crossing, waiting for a ferry to Asiatic Turkey. They were looking for a remote village in which to live, a village within two or three hours' drive over dirt roads to a district law court town.

As they drove down the western coast of Turkey, he was talking again about England and the British Museum. "We could go there together. England is pleasant in the winter. Remember how much you loved Indian restaurants and the London theater scene?"

"I couldn't go to England now," she responded quietly. "This is my first time in Turkey. I just got here after all that studying." He began talking again about leaving her and the child. This talk made her exceptionally sad. Only occasionally did she acquiesce in her husband's view that English scholars belong in the British Museum and not in a muddy Turkish village without electricity or running water.

"Well, just give me something to hang on to; just give me a date when you'll be back," she said.

"How can I tell you when I will return, when I haven't left yet?" he responded.

He would leave her by hitch-hiking a ride on a tangerine truck five days after they moved into the village, one day before his thirty-second birthday, and 13 days before Christmas in 1966. She was devastated. She had not had time to learn to speak Turkish, for she had needed to plan her research to accommodate his sabbatical schedule. Almost no one in the village spoke English. She could not even ask someone where to buy a chicken or a donkey's load of wood. They would have no heat in winter.

As if all nature wept at her abandonment, a huge winter storm swept into the seaside village and for five days and nights she and her young child could not leave the cottage. Waves beat against the cottage's foundation. Rain pounded at the windows and roof. It was cold. It was damp. They had no heat. They lived in their snow suits. Water seeped in under the door and windows! She had to wade through three inches of water on the floor to reach her bed.

Out of the depths of her despair she made a plan . . .

BEING

The rain has stopped. The storm ended. I dragged the wooden table across the cement floor to the wall and set my portable Olivetti on it. I looked around for the kerosene lamp and placed it on the table too, along with several pencils, ballpoint pens, and my notebook. I finally had a desk. I had bought some eight-by-eleven lightweight airmail paper; I inserted two blue sheets and began typing my first impressions of the village: "Small stone houses among fields and orchards. Older style stone houses in the upper village." The village was spectacularly beautiful after the rain stopped, even with its muddy, puddle-filled roads. "My toddler and I walked along the seashore to the ruined and boarded up houses there, abandoned by Greeks decades earlier. Now, as the baby naps, I begin my first anthropological field notes written in Turkey."

But a stranger's voice at the door made me quickly turn. There stood a young Turkish man, beardless, in long pants. "Probably in his early twenties," I thought. He was good looking with straight jet black hair, streaked with gray, worn somewhat longer than most Turkish men his age. He began in halting English, staring at me as I stared back—"I've come over the hills from Bodrum to find you. I heard a young woman and child were living in the village here by the sea and that she didn't speak any Turkish. They said her husband had left the village and the Bodrum area five days ago, but she stayed."

"My name is Adnan," he continued. "I come to see what you need. I know a little English. Can I help you?"

I stared at him, speechless. What did I need? Could he help me? We stared at each other in one of those pregnant moments that are not supposed to arise between a man and a woman who find themselves alone together in a Turkish village.

Gradually, as I stared at him, I began to realize that he could help me. I invited him in for coffee and asked if we could go together to speak to village women—to find one who would look after my son for several hours each day, so I might work unhampered. Several months later at dinner, as I fed my child a spoonful of rice pudding, I turned to Adnan and asked him to tell me a Bodrum proverb. Eyeing me with a worldly smile, he told me two: "The best pears are eaten by the bears," and "Every pot has its cover."

THE BERKELEY REPORT

I will first comment briefly on the report that I quote below. The "Report," written in her stilted academic style of the 1960s, contains facts, allegations, and assumptions that precede the accusations, although, in the younger June's telling of events, she sometimes confuses the personal with the historical. Greece and Turkey did come very close to war over Cyprus in December 1967 and she *was* the only foreigner at that time in the Bodrum region of thirty thousand Turks. It's not hard to imagine why the Turkish authorities ordered her out of the country in December 1967. They thought they were going to war with Greece. And the authorities had no idea why she was living there in the first place.

EXCERPTS FROM THE REPORT TO
THE ANTHROPOLOGY DEPARTMENT AT
THE UNIVERSITY OF CALIFORNIA, BERKELEY
CONCERNING JUNE STARR'S EXPULSION FROM TURKEY
BY THE TURKISH GOVERNMENT IN 1967.[2]

Preperation for Field Research

Before leaving for Turkey, I had mailed a two-page statement of research aims to the Turkish Consulate in Chicago with a covering letter from the Chairman of the Anthropology Department at the University of California, Berkeley. I asked for a year's research permit in order to study two villages' methods of dispute resolution. The area chosen for research was in western Turkey, the Izmir-Muğla region. I asked that all problems relating to living in a village, and keeping a car in Turkey be explained to me. Within ten days I was given a year's Research Visa, allowing me to live in any two villages I chose. The permit was a pink slip affixed in my passport. In a separate letter they said that after four months my car should be put on a *carnet de passage,* or I would need to pay a small car tax, both of which could be arranged in Turkey. The ease with which I obtained the original visa and the clarity of the statement about the car gave no forewarning of what was to come. In terms of the problems I encountered with the Turkish authorities and the kinds of verbal charges they made against me, I suspect that at three points I made mistakes that gave them grounds for mistrusting me. First, I settled into an area near Greek islands where antiquities were exposed on the ground and smuggling was not an infrequent occurrence. Second, I settled in a village on the sea in which lived a foreigner whom they already suspected of illicit activities. And third, I hired a local Bodrum man as a language teacher and research assistant. He was poor, he was known in Socialistic circles, and—perhaps the greatest offense of all—he was clever.

Charge One

"You chose to live in an area of ancient history. You have a Land Rover. Your field assistant, Adnan, has a motor boat. You are either taking things from the sea or taking things from the land. Are you smuggling antiquities out of the country, Mrs. Starr?"

Antiquities exist in the ground all over Turkey. It would be hard to find a 30 mile area that has not at some time been occupied by ancient people and thus filled with their discarded things. Smuggling is a problem all along the Bodrum sea coast, and contrabandists are a fact of life in all her border areas. The things being smuggled differ, of course.

Before settling into the Bodrum area, I spoke to Americans at the United States Consulate in Izmir, who foresaw no problems with my living in a village near Bodrum, especially since I had a year's research permission. After choosing the village of Karakaya, I contacted several of the influential men in Bodrum, including the Kaymakam (the Administrative Director of the Bodrum district). He seemed pleased that I was staying there because some elite Turks in Bodrum wanted to develop Gümüşlük, the seashore area of Karakaya, as a tourist area; therefore, the more foreigners who had lived there the better. No one mentioned trouble with either smuggling or Luke Barnstorm, the Western writer who lived in the village. In fact, the local Turkish authorities seemed relieved that we were planning to live in the same village as Barnstorm, who held both American and German citizenship. Born in Germany, he was writing freelance human interest stories about rural Turkey for German newspapers, or so he said.

In the early period of choosing a region and a village, I did not go to Ankara, the capital of Turkey. Had I done so, I might have established contact with Turkish social scientists who could have advised me on problems of the region and might have sponsored my research, lending me their stature in the eyes of the Turkish government. (Before leaving for Turkey, I had corresponded with both Lloyd Fallers, the leading American anthropologist then working in Turkey, and Nur Yalman, an elite Turkish man who is now a citizen of the United States and a professor of anthropology at Harvard. Professor Yalman was [and is] perhaps the most knowledgeable person in America regarding Turkish village affairs.)

Although hindsight says I should have gone to Ankara to find Turkish social scientists to sponsor my research and intervene for me with the Turkish authorities, given my immediate concerns I had no choice. I was trying to preserve my marriage. I thought the way to do that was to accept my husband's authority and decision-making powers. Hindsight says you can only postpone, you cannot prevent. But, at that time, I had no perspective on my relationship with him.

Charge Two

"You live in a village where a German man whom we know is a spy for Greece or for Germany or the United States lived. We know he is a spy because we expelled him from Turkey for spying. . . . You live in a house constructed by that man's friend, an Englishman named Hays. You and the spy were engaged in smuggling. You are either bringing in contraband from Greece or sending antiquities out of Turkey. Are you digging antiquities from the land, Mrs. Starr?"

The House

By the end of November 1966, I had visited, near Izmir, several villages that I liked. The people were friendly and interested in me and my little son. The villages were between 400 and 1,000 in population (the size requirements I had set for my comparative study of disputes resolved in small and in large villages), but no empty house was available. Everywhere, villagers said, "It will take three months to build a house, and it can't be started until the spring comes in February or March." In the Bodrum district I visited about ten villages before deciding on Karakaya. A few of them had an empty house or two, but they were incredibly small—only one small, eight-by-ten-feet room, eight feet high. All were in disrepair.

During the time I was searching for a village, I assumed that my husband would stay throughout my field research because I had become so compliant. Since he was still in the picture, I decided not to rent a room from a Turkish family; even the largest Turkish village houses in the region had, at most, three rooms. Furthermore, all Turkish village household members eat and sleep in the same room, stacking their bedding and mattresses against the wall during the day. My renting a room from a family would have cramped their usual activities. Finally, my adorable son was not as docile and controllable as Turkish two-year-olds. In middle-class America, we speak of the "unbearable twos." He would have been a constant bother to a Turkish family and a source of anxiety for me in my relationship to them. To live in a self-contained unit with my family seemed my only option.

There was an unoccupied furnished house in Gümüşlük that had been built as a summer cottage by Commander Hays, a retired Englishman. It had two bedrooms, a large living room, and a separate kitchen. It had one Western-style toilet, which I considered a great boon. Finally, the village *muhtar* (headman) and nearby neighbors seemed pleased that we might live there.

Never in my wildest dreams had I planned to live alone with my child in a Turkish village. Luke Barnstorm, a 53-year-old Austrian by birth, American by citizenship, and journalist by profession, had been living in Gümüşlük for nine months. He spoke fluent Turkish and seemed to have good relations with both the villagers and the administrators in Bodrum, where he had lived for three years prior to coming to the village. I realized there might be problems in associating with an expatriate, but I thought, "I'll leave the village, if difficulties occur."

My husband's announcement that he would soon be leaving led me to settle in this village and Commander Hays's empty house. Without my husband, child care loomed as an even larger problem. I felt desperate when thinking about how I might cope with the situation, but realized that even several weeks' respite there would allow me to begin speaking Turkish and start adjusting to the rhythm of Turkish village life. In four or five months I could learn enough Turkish and Turkish behavioral norms to move to another village where there were no other foreigners and thus function entirely on my own. But, at that moment of choice, I could not even order firewood, much less ask a villager to kill a chicken for me.

Charge Three

"Do you know Adnan? How long have you known him? You have known him four months, we have know him four years. We know he is a bad man. Do you want to argue with us about that?"[3]

After meeting Adnan I hired him to be my field assistant. He was a professional sponge diver, which is a seasonal occupation. In winter, he was unemployed. He knew a little English, asked intelligent questions about my work, knew the people in Karakaya and Bodrum, and seemed to get on well with everyone. Originally, I had thought to hire a Turkish graduate student in anthropology, but I soon learned that there would be problems in taking an urban Turk to live in a village without electricity and running water. In those days, urban Turks considered the villagers dirty, backward, and illiterate. So it seemed easier to have a clever local man working with me. I could teach him to be an assistant. From Adnan I could learn the local speech styles and local vocabulary of the people whom I was studying. Adnan was a good assistant, an excellent source of information, and a true friend. Initially, he provided us with his kinship network, which, because of my association with him, welcomed me into their group.

I am sure that Adnan was not and had not engaged in smuggling or in selling antiquities to foreigners. First, he was one of the poorest men in Bodrum. Had he been a smuggler, he would have had cash all of the time.

The Secret Intelligence Service seemed aware of Adnan's socialist sympathies. In a country where about four percent of the population owned 70 percent of the industry and received 38 percent of the Gross National Profit, socialism was a legitimate threat to those in power, even though at that time the Socialist party was doing not badly in elections.

Adnan had been a close associate of Barnstorm's before he became my friend. It was this linkage between Adnan, Barnstorm, the village of Karakaya, and me that the Turkish authorities would refer to again and again.

"TELL JUNE TO BE CAREFUL. LUKE BARNSTORM IS GOING TO BE SENT OUT OF THE COUNTRY." —FAMOUS BODRUM SMUGGLER

The above message reached me on a mild spring evening in early April 1967. Barnstorm was not to find out for several more days. Actual charges were never explicitly made against him, but one ministry refused to renew his press card, allowing the Ministry of Interior to deny him a residency permit in the absence of permission to work in Turkey. He was ordered out of the country as *persona non grata* and left with his dog Pluto early in May 1967.

Barnstorm's plight worried me, for I thought that Adnan and I would be irrevocably implicated in his activities. Also, I had not received official responses to my requests to live in a Turkish village, to have a Turkish residence permit, to see Bodrum court records, or to put my car on a *carnet de passage*. To check on these permissions, I went to Ankara near the end of May with Adnan, my child, and a thirteen-year-old girl from Gümüşlük to care for him. Adnan had friends in Ankara who might help me, and he would help with the two-and-a-half days' drive. My Turkish by now was adequate, but not excellent; having a translator for the ministries seemed a good idea.

We packed the Land Rover and spent the night of May 21, 1967, in Bodrum town, all staying at Adnan's mother's house, a one room stone cottage. We planned to leave for Ankara the next morning.

Dynamite in the Car

A night in Bodrum was a festive event in our lives then. A party was planned at one of Bodrum's eating places. Adnan and I, Kathy Flegal—an undergraduate anthropology major from the University of California at Berkeley—and local Turkish people attended. As we returned to Adnan's mother's house at about 12:30 in the morning, we were followed by two Bodrum policemen. I went to bed. Adnan went to look after my Land Rover. He found a policeman named Osman Bey[4] in it. Osman asked for the keys to my two locked suitcases, saying he wanted to make "control." Adnan woke me. I dressed and, arriving at the car with the keys, said in Turkish, "I have nothing to hide, but you had better get permission to search my car." The policemen answered that he did not need permission. "I will give permission, but I want some witnesses," I said. We went to the veterinarian's house—he was a friend of mine and known to be one of the most powerful men in Bodrum because of his political party connections. We also woke up the assistant director of the Bodrum Museum, a man named Yüxel. He was an acquaintance, had been at the party, and spoke fluent English. He was said to be fanatical about antiquities being smuggled out of the country. Barnstorm thought that Yüxel was behind his deportation.

While waiting for Yüxel to dress, I said to Osman: "This is no way to treat a foreigner, and someone who is a guest in your country. Why couldn't you search my car in the morning when it is light?" (There had been other stories of police bothering foreign women at night.) Osman answered, "We are going to make control now." Not having the slightest foreboding of what was to come, I said, "You will be sorry for treating me this way. I am a very important person. I come from the biggest University in America to study the people in Bodrum and Gumusluk. I have many Turkish friends. My American University is very powerful. I am going to Ankara tomorrow. I will tell them how you have treated me like a criminal."

Permission to search the Land Rover was given by the Public Prosecutor at 4:30 A.M. The police went through every basket and suitcase. Under the last suitcase they found fisherman's dynamite, which I mistakenly recognized as only child's clay. My assistant and I were arrested, held prisoner at the police station, and at 9:00 A.M. marched through town with a police escort to the courthouse, where the state opened a case against us. Our trial took two days, at the end of which we were found innocent, on the grounds that my car had not been locked—anyone could have planted the dynamite there. The judge reasoned that if we had been carrying dynamite about we would have placed it in one of my locked boxes.

Analysis of the Dynamite Frame-Up

Both Adnan and I thought that Osman had planted the dynamite in my car. Adnan actually saw him *inside* my Land Rover, so he had opportunity. He also had motive—to discredit me so that any reporting I did concerning his oppressive behavior would be undermined by my "criminal" actions. Possessing dynamite would have also justified his searching my car. (In Turkey, as in most countries, it is a crime to possess dynamite without a license.)

Under oath, Osman swore in court that he had not entered my car before I was present. We take his lying over this incident as proof that he thought he had something to hide. As he said, he was free to do whatever he thought necessary to "make control." The real question was at what point did Osman put the dynamite in my car? Did he do it in response to my threatening him with reprisals from Ankara, or earlier? Whenever he planted the dynamite, the fact of dynamite in my Land Rover proved that he was right to suspect us of something, for although there were no antiquities and no gold in my car, something forbidden by law had been concealed there. Another question concerns who decided my car should be searched in the first place, and did the person who suggested the search plant the dynamite?

For the last months I have favored the following explanation: Yüxel, the Turkish patriot and director of the Bodrum Museum, may have thought I was carrying antiquities to Ankara and, learning at the party that we were on our way to Ankara, he told the police to search my car. He did leave the restaurant for an hour during the evening, saying to me he had to meet a certain lady. Sometime during the course of our dinner conversation he asked me where I had parked the car (parking being a problem on Bodrum's medieval, cobblestone streets). He therefore knew that my car was on a back street near Adnan's mother's home, and he knew I was leaving for Ankara in the morning. I do not actually believe he planted the dynamite himself, because I do not think he necessarily was intent on getting rid of me or scaring me away. I think he merely wanted the car stopped and searched before we left for Ankara. Did he provoke the police into a search, and did my high-handed attitude toward Osman Bey provoke him to plant dynamite to save his reputation? Despite the fact that the Bodrum district court found us innocent, the Ankara officials would always wonder why I was carrying dynamite in my car.

The Bodrum judge who heard the case had another theory, which he wrote in his court decision exonerating us in the dynamite case. He said someone in Bodrum was jealous of the fact that Adnan was my assistant. They were jealous that I, a mere female, drove a red Land Rover around town and lived the life of a "free woman." Thus, the dynamite had been planted so that either I would become frightened and leave the area, or Adnan and I would be found guilty of carrying explosives and would spend up to six months in jail. At the time, I considered this explanation too far-fetched; but later, after analyzing dispute cases I had collected in the villages, I found that *envy* often does play a large role in rural Turkish life.

ANKARA TRIP I (MAY 23–JUNE 10, 1967)

Once in Ankara, Adnan and I hung around government offices until someone made an appointment for us at Şübe (Department) 4 of the Ministry of the Interior, where we could acquire a residence permit. About ten days after we arrived, I found myself waiting anxiously outside a locked door with Adnan, hoping that the assistant director of the department would see me. He angrily opened the interview: "Why did you go to your Embassy and have them write to us about you? You should not have done that. That embarrasses us. We would have given you a permit, but now that you have gone to them, it is no longer easy. We are quite angry with you."

When I asked why my permit to live in Karakaya had not been granted, he asked how I had happened to settle there. After I told him about the three-room empty house, he stated it was forbidden for me to continue living in a Turkish village. I asked why.

He answered, "Why does a young woman want to live in a village when our Turkish cities are so fascinating? Why does an educated young woman, an intellectual, hang out with an illiterate fisherman, when we have such interesting cultured Turkish men in cities?"

I explained the nature of my anthropological work.

He answered, "Do you know Peter Benedict?"

Of course, I had heard of Peter Benedict. He was one of two students that Lloyd Fallers had brought to Turkey with him from the University of Chicago. Benedict was supposed to be working in Yatağan, a Turkish town about four hours from my site. I had mailed him a note, but never heard back from him.

The assistant director said: "He wanted to go to Yatağan, but we forbid that."

"Where did he go?" I asked.

"He will not go to a village," he replied darkly. "He may leave the country. And as for you, you cannot live in Karakaya," he said. "We forbid it."

We returned to Bodrum on June 11, 1967, and I went to see the Kaymakam to seek an agreement about my research in Gümüşlük. I was six months into a village study and was not ready to give up the idea of writing a legal ethnography of the processes of social control in Turkish villages that allow villagers to settle conflicts among themselves. Furthermore, I had gone through much to get established in the village, both emotionally and psychically, and I did not want to start over in another part of the country (as the Turkish officials wanted).

The Kaymakam and I discussed the situation and he agreed to my plan. I would rent a house in Bodrum town and live there four days a week; three days a week I would stay with friends in Karakaya. I would move out of the house I had rented in the village and rent a one-room house for typing and interviewing. I rented a house and began planning my move.

Several days later the Jandarma[4] commander, who resided in Bodrum town, came with his entourage to the seaside cafe in Gümüşlük. It was June 14, 1967, and a lovely sunny day. Between the sun and the drinks, the commander became very drunk. Some of the male villagers gathered around him and told him of my troubles about staying in the village. He roared, "If she doesn't have written permission from the Kaymakam, she cannot stay in this village, even overnight." He instructed the village Jandarma commander to inform me of his decision, which he did promptly. The next morning my son and I moved out of the village with all of our belongings.

I spent the next two weeks in Bodrum town looking for a house to rent while we resided at a small hotel on a cobblestone street by the sea. In this same period, I received a letter from the Ministry of Justice. The dear elderly judges of the Yüksek Hakimler Kurulu (the Supreme Council of Judges) with whom I talked about Ataturk and the Turkish legal revolution had given me permission to study *any* court in Turkey, including the Bodrum law court and her records. I was sad and depressed at having to abandon my village study, which I had prepared for during my graduate work in Berkeley's Anthropology department, but I was also partially elated: I had found some Turks who believed me and thought my project was valuable.

With the judges' permission I could design a full-time research project for studying the civil and criminal courts of Bodrum town. Luckily, I had been watching certain cases that villagers from Karakaya had taken to court; I had been spending about one day a week in court since February 1967. My assistant and I had established good rapport with the two Turkish judges and the public prosecutor. By July 1, 1967, about six months after I began my court observations, I had written out a plan of research on the court that would take approximately one year. At that time we were working at the court or at home on court data about eight hours a day, five-and-a-half days a week.

On July 12 I was called to the police station. Naturally, I was scared and worried. Osman Bey tried to make me sign a paper stating that I promised not to conduct research in either the village of Karakaya or in the town of Bodrum. I refused, saying I knew my village research was forbidden and I had given them my word not to pursue that. However, I did have permission from the Yüsek Hakimler Kurulu to study the Bodrum law court, and I was undertaking that research. Some of my Turkish friends felt I should have signed this paper, that in refusing I was being "unduly legalistic." The same day I went with Adnan to see the Kaymakam to tell him of these events; he, too, thought I should have signed the paper. "It is unimportant," he said. "You could have signed it."

Two days later I was again called to the police station where Osman Bey read me an order which had come from Ankara several days earlier. This order said, "Since we have forbidden her village study, she has no longer a reason to stay in Turkey. Therefore, she must leave the country by the end of the month." Had I signed the paper Osman submitted to me, this order would have been easy to enforce.

But I had followed my instincts and not taken the advice of my Turk-ish friends. The Turks have their own view of their government and bureau-cratic system, which differs from mine. Turks, for the most part, go along with the bureaucracy as much as possible. If they mistakenly sign something and it turns out to be important, then they find someone equally important to "fix" the situation for them. I, however, had used up a lot of "social capi-tal" in dealing with the Ministry of Interior Affairs in Ankara and with the local police. Without many important connections and favors to call in, I had thought it wise to be more cautious.

ANKARA TRIP II (JULY 15–25, 1967)

That same night, Adnan and I telephoned Oktay, a friend in Ankara and a reporter on one of the largest and most influential Turkish newspapers, the *Cumhuriyet*. He had become a friend of Adnan's while vacationing in Bo-drum in summers. Oktay had heard about me and my earlier problems, and he wanted to help. Partly, I think, he wanted an international love story to write about in his newspaper, and he thought ours would do. But, we did not know that then. He called back the next day, claiming to have seen my gov-ernment file. "You must come to Ankara immediately," he said, "and you must come alone. I will expect you to be in Ankara by the following evening."

I left Bodrum on the 5:00 A.M. bus for Ankara.

Oktay said the evidence in my dossier was as follows: (1) I had known Luke Barnstorm, whom the Turks had put out of Turkey for spying and/or smuggling, (2) my assistant had also known him, (3) I had a Land Rover, (4) I had lent some money to my assistant to make a payment on his sponge-diving boat, and (5) we had been arrested carrying dynamite in my car. Oktay thought we could clarify the issues if I could meet with the director of Department 4 of the Turkish Ministry of Interior Affairs. Through Oktay's newspaper's connections to an important member of the Turkish parliament, an interview was arranged. When I was finally shown into the director's office, three or four armed military aids were standing there hold-ing three different dossiers. One turned out to be mine, one belonged to Barnstorm, and the third to Adnan. Mine was by far the thickest, perhaps three-and-one-half inches thick. One soldier stood at attention with a rifle and bayonet held at his waist. I was the only woman in the room.

The director, Ismail Bey, sat opposite me; Oktay sat next to me, ready to translate if needed. Ismail was six-and-a-half feet tall—a huge man in dark sunglasses. No one spoke in English. However, Oktay understood my Turk-ish and would rephrase if need be. I was intimidated.

Ismail began by asking me rhetorical questions. He was not interested in my answers, only in making charges.[6] When I asked what kinds of things I had been smuggling, he treated me as though I was being facetious. They asked how long it would take me to finish my work in Bodrum.

"One year," I said.

"Impossible," he said. "We will let you stay one more month in Bodrum and one year in a village or market town near Ankara. That way we can check up on you. We can make control."

I had witnessed how Turkish police behaved under normal circumstances around Bodrum. If police in a different region were told to keep track of me, who knows what might happen. How would villagers react if every week an entourage of police arrived to ask questions about me? No thanks.

Even though I no longer felt secure in Bodrum town, at least I had friends there, a work schedule, and an idea of how the legal and organizational structure operated in the Bodrum region. Villages around Ankara are totally different from those in southwestern Turkey, probably settled by different people with different customs. And the villages near Ankara had been studied by the two great anthropologists of Turkey—Paul Stirling from England and a Turk, Ibrahim Yasa. Perhaps I was narrow minded, but I did not want to live in a mud house in a poor wheat growing village on the Anatolian plain, and I did not want to begin anew my Turkish research project! To top it all off, the director said I could take neither the village girl who looked after my child nor my field assistant—the two people I most trusted—with me to Anatolia.

After a week of meetings we finally reached a compromise. I could stay three more months in Bodrum, if I would then leave the country. I would have agreed to almost any proposal if it meant I could stay longer in quaint, medieval Bodrum by the sea, which I had come to love.

I was issued a residence permit for August 17 through November 17, 1967. The United States Embassy in Turkey had written a strong letter to the Turkish Ministry of Foreign Affairs on my behalf. Professor Laura Nader, of the Anthropology department at University of California, had sent telegrams to the minister of the Interior, the Bodrum Kaymakam, and the Muğla Vali.

We returned to Bodrum, and a peaceful routine took over our days. The police left me alone. Slowly, the personnel at the court became friendlier and began to let us see case dossiers—first, the records of cases we had observed each day. Later, we hand-copied the court statistics on cases heard and finalized each year.

At the end of October 1967, a letter came from the Muğla Vali asking how much longer I needed to finish my research. I asked for three more months, hoping that three might be a magic number in Turkey. On November 12, a message came to the police station (I was not to learn of this until November 20th) that Department 4 was preparing a new three-months residence permit for me. I was asked to give my passport to the police so that it could be made ready. When I asked for a three-month extension, I wrote to one of the under directors of Department 4, telling him how well my work was going and asking for more time. He had been influential in arguing for my first three-month residence permit.

On December 6, 1967, I was again called to the police station. Adnan and I put off going because we were just finishing up a large project at the court that involved hand-copying court records.[7] At about 11:00 A.M., with the job completed, we went to Bodrum's harbor, where the police station was located. The chief of Police told me they had received a telegram that morning saying that Department 4 had decided not to renew my residence permit; they therefore requested that I leave the country in one day. We asked to see, and were shown, a copy of the telegram, which did come from Ankara and appeared to say I had to leave immediately. I said, "If I got into my Land Rover and stopped only to pick up my infant son, we could not in one day reach a border through which I might drive a car.[8]

By mid-afternoon the order had been amended to read "must leave Bodrum in two days," and I was asked to state if I would leave from Izmir or Istanbul. I choose Izmir—it was closer to Bodrum.

In truth, I had already been packing because I was very concerned about the flare up in trouble between Greece and Turkey over Cyprus, which had considerably worsened in the last ten days. The Turkish military had been moving cannons through Bodrum town at night to the seaside areas closest to Greece. I did not want to remain in Turkey if they went to war with Greece, and I had been getting ready to leave at an hour's notice. At this time I was the only foreigner in the entire Bodrum region of 30,000 people. Therefore, the idea of Christmas in the United States with my parents in Cincinnati did not seem like a bad idea at all. I was emotionally and psychologically ready to leave for a while.

The next morning, we left in the Land Rover—my son, Adnan, and my policeman escort. When we ate, the policeman ate, and I paid for everyone. He was a very young man and sorry to be caught up in this situation; after we left town, he switched to civilian clothes, so I felt less like a criminal.

The police escort turned me over to the director of the Izmir police, who took my passport without a trace of humor. To him I was clearly a wicked and dangerous woman, guilty of some unimaginable deeds against Turkey. He warned me that I must leave the country with all haste and asked what I intended to do next.

"I shall go to the American Consulate in Izmir," I said.

"Good," he answered, "tell them to send someone over to pick up your passport. They will be responsible to see that you leave the country." The American Consulate treated me quite well, except he seemed a little disappointed that I was not more like Mata Hari.

Turkey had an attitude about foreigners who bring cars into the country. The law is that the car cannot be sold to someone in Turkey and it cannot be left in Turkey. The Land Rover had to leave with me. I found a Turkish Customs broker, who, with the usual delays and confusion that I had learned to expect in dealing with administrative matters in Turkey, managed to get the car off my passport and into Turkish customs to ship it to me in New York City.

DEPARTURE

And so, after almost a year to the day that my husband left me, my son and I left Izmir, Turkey, for Athens. It was December 10, 1967, and we had become totally different people from a year before when we first arrived in Turkey. My son was bilingual in Turkish and English, preferring to speak in Turkish. His Turkish was as good as a native child's, and he had no accent. I too spoke Turkish and well enough, although I never entirely mastered the elite forms of address. I spoke like the rural villagers among whom I had learned the language. Their accent was my accent, which was okay with me.

Yet, for me the transformation was even greater than being able to communicate with everyone in Turkish. I had crossed over the threshold into adulthood and independence. I had filled reams and reams of paper with field research data, and I had no doubt that eventually the data could be transformed into an anthropological book. I no longer questioned whether I had the "right stuff" to become an anthropologist. For better or worse, I was one. And, like all other anthropologists, I had many stories to tell about how I survived in the field.

Was I glad to be leaving Turkey? Probably not. I had made a new life in Turkey, one I quite valued. I was devoted to Adnan and to other friends. The Report to the Berkeley Anthropology Department bears witness to this as it concludes with a detailed plan of what I wanted to study in the coming summer of 1968, when, I hoped, my child and I would fly in on a tourist visa for a three-month visit. In fact, we made that return field trip,

only to be discovered after a-month-and-a-half in Bodrum, when I was
again ordered out of Bodrum. That final month in Turkey I traveled all over
the country with my son and Adnan, gaining comparative knowledge of
other small seaside towns and other regions of the nation.

*Perhaps, the only way to end this saga, covering huge distances in time, space,
and human psychology is to quote from the conclusion of Starr's Report to the
Berkeley faculty. Writing about her expulsion from Turkey, June states:* "The interesting thing is not that I was finally denied a residence permit and ordered
to leave the country, but that with all my initial mistakes I was able to forestall the logical conclusion as long as I did."[9] *She had managed to stay thir-
teen months. She had circumvented the stigma of being branded a criminal or a
spy. What she did not realize was that it would take her 30 years before she was
ready to tell "the true story," "the whole story" of what had happened. And this
isn't all of it yet.*

NOTES

1. The author wishes to thank Ron Krotoszynski, David Papke, and Florence
    Roisman, all of Indiana University School of Law, Indianapolis, for their use-
    ful comments and criticism.
2. The report was originally called "Real and Unreal in the Context of Suspi-
    cion," dated April 16, 1968. Hereafter, I refer to it as *The Report.*
3. Interview with Ismail Bey, Director of Secret Intelligence Service, charged
    with observing and keeping track of resident foreigners in Turkey (Director
    of Şübe 4, of the Turkish Ministry of Interior Affairs) July 1967. Compiled
    and edited from Fieldnotes July 1967, and "The Report to Thesis advisers at
    the Anthropology Department, Berkeley" (April 16, 1968).
4. In Turkish it is common to address men by their first name followed by the
    honorific "*bey,*" which is similar to the use of the term "*san*" in Japanese and
    translates to "gentleman" or "sir."
5. Based on the French gendarmes system, *jandarma* are members of the Turk-
    ish national police organization constituting a branch of the Turkish armed
    forces responsible for, among other things, general law enforcement in rural
    districts. Their commanders are career officers. Bodrum had urban police for
    Bodrum town and the headquarters of the district's Jandarma.
6. See charges I, II, and III, pp. 10–11.
7. The Bodrum judges and court personnel never agreed to let me photograph
    records although eventually, in November and December 1967, they allowed

me to bring a movie camera into the courthouse to film several cases. See Adliye. *The Ethnography of a Rural Turkish Law Court.*

8. In summer, car ferries operated between Marmaris in Turkey and Rhodes in Greece. Also there was a car ferry from Turkey to the Greek island of Samos, but in winter, because of unreliable weather, the ferries did not operate.

9. Starr, June. Op. cit. ft. 2.

BIBLIOGRAPHY

Starr, June
1985   Adliye: The Ethnography of a Rural Turkish Law Court. Color Video, Film Script, and Teacher's Guide. 30 minutes, sound.
Stirling, Paul
1965   Turkish Village. New York: Wiley and Sons.
Yasa, Ibrahim
1957   Hasanoğlan: Socio-Economic Structure of a Turkish Village. Ankara: Yeni Matbaa.

CHAPTER 4

WILD POWER IN POST-MILITARY BRAZIL

Daniel T. Linger

Many questions were troubling the explorer, but at the sight of the prisoner
he asked only: "Does he know his sentence?" "No," said the officer, eager to
go on with his exposition, but the explorer interrupted him: "He doesn't know
the sentence that has been passed on him?" "No," said the officer again, paus-
ing a moment as if to let the explorer elaborate his question, and then said:
"There would be no point in telling him. He'll learn it on his body."
—Kafka, "In the Penal Colony" (1971 [1919]:144–45)

WILD POWER

IN 1985, CIVILIANS FINALLY ASSUMED CONTROL of the Brazilian government
after more than two decades of military command. Like their counterparts
elsewhere in the Southern Cone, the Brazilian generals, crusading in the
name of anti-communism, had governed through arbitrary decrees and had
sponsored outrageous violations of human rights (Arquidiocese de São Paulo
1985). Repression peaked in the middle years of the regime—the early
1970s—a dark time followed by an agonizing period of controlled disten-
tion (*distensão*) and opening (*abertura*) culminating in the peaceful 1985
transition. The national government has remained in civilian hands since
then, and without question the political atmosphere has dramatically im-
proved. Brazil is now a functioning democracy with elections, a new constitution, free speech, and a far-flung galaxy of political and social movements

(Alvarez 1990, Sader and Silverstein 1991, Andrews 1991, Alvarez and Escobar 1992, Keck 1992, Abers 1996).

The transition, though halting, protracted, and uncertain, fueled hopes for the establishment in Brazil of the rule of law and respect for human rights. Those hopes have been only partially fulfilled. As anthropologists know well, attempts at social change often run into formidable cultural resistance. Culture does not, of course, mechanically reproduce itself. But it is a mistake, as Antonio Gramsci (1971) recognized, to assume that cultural formations are *necessarily* plastic or evanescent. Some are extraordinarily tenacious. This paper argues that *wild power,* the cardinal repressive practice of the dictatorship, survives in *zones of transgression* located on the margins of the Brazilian state's authority and in the recesses of everyday Brazilian life.

Wild power is a form of coercion that is unregulated, unofficial, unpredictable, potentially annihilating, and therefore terrifying. As I will show, wild power also has connotations of invasion, penetration, and feminization, drawing on and reinforcing stratified concepts of gender and violent images of sexuality. Zones of transgression, the interactional spaces in which wild power operates, by definition lack legal or bureaucratic regulation, leaving victims defenseless against the machinations and caprices of their persecutors. Wild power is both a practice and a cultural object, a complex of ideas and feelings. Power, exercised for good or ill, is a signal feature of hierarchical relationships in Brazil. Although power can be beneficial, a resource in times of need—for example, the power of God and other sacred entities, or of well-connected patrons—it can also (in its negative, wild form) be dangerous, manifesting itself in arbitrariness, abuse, and explosive brutality. The practice of wild power enhances its cultural salience; its cultural salience permits powerful actors to use local forms of fear and uncertainty as political weapons.

Sometimes wild power is lethal, as it was, often, during the dictatorship and can be, still, in the the streets, jails, and police-station basements of post-military Brazil. Murders of homeless children, prison massacres, vigilantism, and extrajudicial executions have all received wide coverage in recent publications (Amnesty International 1990, 1993a, 1994; Americas Watch 1987; Chevigny 1990, 1995, 1999; Dimenstein 1991; Pinheiro 1983, 2000). But wild power thrives principally in the shadow of extreme violence. That is, the effectiveness of wild power rests upon the perceived possibility of such violence and only indirectly, therefore, on its realization. One need not kill often to create an atmosphere of menace and intimidation.

In this chapter I explore gang rapes and terror-squad abductions as they occurred during the 1980s and early 1990s in São Luís, a city of 800,000 and capital of the state of Maranhão in northeastern Brazil.[1] Although no one died in the cases discussed, the victims suffered violation, humiliation, and terrorization. The acts themselves were elaborate death threats, for victims did not know how far the violence might go. Gang rapes, known in São Luís as *maratonas* (marathons) or *curras* (corrallings), and terror-squad torture sessions, commonly called *seqüestros* (kidnappings), usually take place at night in secluded areas such as abandoned houses, distant beaches, or patches of bush at the edge of the city. The victim of the maratona, almost always a young woman, is raped by each of her abductors as the others look on and keep an eye out for the police or passersby. The victim of the seqüestro, usually a young man, is tortured by the kidnappers, who have nothing to fear from the police or passersby because the kidnappers are, by all indications, *themselves* members of the police force.

Although carried out by disparate parties in dissimilar circumstances, both maratonas and seqüestros reproduce haphazardly and on a small scale the most abhorrent institutionalized practices of the dictatorship. Wild power was the defining feature of the military's torture centers, which trumpeted themselves as ultimate zones of transgression. One former prisoner interviewed by the Brazil: Never Again Project, José Elpídio Cavalcanti, recalls the apt but chilling salutation he received upon his arrival at an interrogation camp outside Fortaleza in 1974: "THIS ISN'T THE ARMY, THE NAVY, OR THE AIR FORCE. THIS IS HELL" (Arquidiocese de São Paulo 1985:240, capitals in original).[2]

Hence, despite the relaxation and then demise of blatantly oppressive military rule in Brasília during the 1980s, wild power continues to be propagated in events such as maratonas and seqüestros.[3] An obvious danger is that future authoritarian rulers might be able once again to mobilize wild power systematically and on a large scale. But I wish to focus on a more general and immediate issue. To the degree that wild power remains embedded in common sense as an expected, "natural" mode of operation of hierarchy and repression, it authorizes and normalizes a wide range of abusive practices that are not limited to those carried out by agents of the state. Significantly, although during the 1970s the military regime refined the use of wild power into a diabolical high art and therefore deserves special and severe condemnation, then and now no class or elite group holds a monopoly on it. For example, often the perpetrators of present-day maratonas are (according to police, press, and word-of-mouth reports) working-class or unemployed young men. The persistence of zones of transgression throughout the social hierarchy is, I suggest, ipso facto a serious political pathology, whether or not wild power once again entrenches itself as high-level government policy.

The discussion below makes use of materials I collected during two field trips to São Luís (1984–86 and 1991), including police records of maratonas, newspaper reports, conversations with perpetrators of and witnesses to maratonas, and an interview with the victim of a seqüestro. Working with such materials presents significant obstacles to an ethnographer. Police and court records are hard to obtain; as is often the case in Brazil, access requires the intervention of a sympathetic party and typically has certain conditions. I was able to contact some civilian perpetrators, but only through trusted intermediaries, and only after a long stay in the city had established my credentials and my discretion. As for police perpetrators—I did not even attempt to interview them. In the first place, I doubted whether such conversations would be revealing. But more importantly, I did not want officials to know that I was looking into matters of police brutality, in part because such knowledge could have jeopardized my research, but mainly because I did not want to endanger my family or myself. After all, threats and terror tactics were the basic repertoire of these people. I did find it possible to interview some victims, conditioned on assurances of absolute anonymity.

I discuss the culture of power revealed in the ethnographic case studies. I close with an interpretation of "Geni and the Zeppelin," a popular song by the Brazilian composer Chico Buarque, which recapitulates in brilliant, concise, and compelling images my own analysis and critique. Chico's song is a sharply ironic condemnation of wild power—a metacultural response, albeit a pessimistic one, to a tenacious and pressing cultural problem.

MARATONAS: "DON'T SHOUT OR YOU'LL DIE"

Women's Views: Maratona as Violent Attack

The *Delegacia Especial da Mulher* (DEM), established in December 1988 and located in central São Luís, is the police station that handles most violent crimes against women.[4] I arrived there one day in 1991 carrying an introduction from the *Grupo de Mulheres* of São Luís, a local women's rights organization. In response to my request, a police officer of the DEM located six relatively complete files—including depositions from victims and accused, and in some cases medical examinations—on 1990 gang rapes in the city.[5] I was permitted to copy these files into a notebook on the condition that I use only pseudonyms for the victims.

Women find it less intimidating to report incidents of rape to the female officers of the DEM than, as before 1988, to male officers at the ordinary police stations. Nevertheless, victims have strong incentives not to bring rape charges: to be raped is shameful, to be assaulted multiply is worse; the men almost always warn their victims, with graphic threats, not to go to the police. There is little motive to press false charges, given the unpleasant, at times humiliating, medical and legal procedures to which an accuser must submit (Ardaillon and Debert 1987). For these reasons, I think the women's depositions are highly credible.

The attack suffered by Carmen is in most respects paradigmatic of gang rapes in São Luís, although some rapes can be much more violent.[6] I have summarized the deposition she gave to the police:

> Carmen, 23, a cashier, was returning from a party at 1:30 A.M. with a woman friend. As three men approached them, her friend ran away and Carmen was seized by a man who, holding a knife to her throat, warned her, "Don't shout, or you'll die." The men slashed her and took her to the mangrove swamp near the Merck Pharmaceuticals plant, where they threw her to the ground among stones and tree stumps. There the three ordered her to undress and raped her, orally, anally, and vaginally. One said, "Let's kill her," but another disagreed when she told them she had three children to raise without a father. One then slapped her face three times, saying that if she went to the police he'd kill her as soon as he got out of jail. Twenty-five minutes after they left she went to the Merck plant to seek help.

Drawing on this case and others, I would suggest that the prototypical maratona takes the following form:

- A woman is walking alone (or with a female companion) on the streets late at night. A group of armed young men accost her and force her to accompany them.
- The men take the victim to a deserted spot. This may be an abandoned house in an empty street or at the edge of an illegal settlement (invasão), or they may take her to an uninhabited locale—a beach, a swamp, a stretch of bush—on the margins of town.
- The woman is forced to undress; the men then take turns raping her at knifepoint (or gunpoint). Often (though not in this case), those men who are reluctant to participate are made to do so by the others. The attacks usually involve anal or oral sex (common but "dirty" sexual practices known as sacanagem[7]) in addition to genital intercourse. These acts may be accompanied by verbal insults (for example, "whore" [puta], "tramp" [vagabunda], or some other accusation indicating promiscuous behavior), beatings, other physical assaults, and/or death threats.[8]

4.1. A *delegada* (station chief) questions a man at the women's police station in Salvador, Bahia (photo by Sarah Hautzinger).

4.2. Women waiting to file complaints at the women's police station in Salvador, Bahia (photo by Sarah Hautzinger).

- The victim—nude, disheveled, often bleeding—is abandoned at the scene of the crime and must seek out someone, inevitably a stranger, to help her.

Gang rapes do vary. Sometimes the assailants use a car, as in the incident described by Jair below. Although availability of a car is not necessarily a sign of wealth—drivers have access to the cars of their employers, for instance—in some cases the rapists are middle- or upper-middle-class men.[9] In one of the DEM cases, the rape occurred during the daytime. In another, a menstruating woman was not raped; menstruation, a sign of pollution and sickness, discourages men from sexual intercourse (Parker 1991:40). In yet another instance, a woman's male companion was taken along with her and was himself threatened with rape (although the assailants agreed finally that "this was not a good idea," perhaps because they thought assaulting a man might have more dire consequences than assaulting a woman).[10] But the rape of Carmen is a prime example of a maratona, its main features replicated in oral narratives, newspaper records, and police files.

Men's Views: Maratona as Pleasure and Punishment

The following narrators, identified herein by pseudonyms, have never been legally accused of rape, although in some cases they have participated in such acts.[11] They spoke with my research assistant Carlos Ramos, who, at my request, collected stories of maratonas during my 1984–86 field trip to São Luís. The words are those of the narrators. First Jair, 19 years old and a secondary school student, recounts a story told to him by a friend, Raimundo, as they were drinking in a bar:

> Raimundo was working in the Lusitana [a local supermarket] and left work about 6:00 for the bus stop. This friend of his Roberto passed by in a car; they're both heavy characters. Roberto invited him to do some drinking and smoke some marijuana. So Raimundo said they smoked a few joints in the car, and then they started drinking beer, then they smoked some more. By then it was 3:00 A.M. They were coming from the beach, and they offered two girls a ride. So they picked them up and went out to Raposa [a remote fishing village] to rape [currar] them, one screwing one and the other screwing the other. Then they switched off. One of the girls ran away, and Raimundo's friend shot at her twice, but he didn't kill her; she escaped. So then there was only one left. They raped that one. Later she said they didn't have to [force her,] that she'd even smoke marijuana with them. I don't know if she said this so she wouldn't have some kind of feud [atrito] with them. He told me there's nothing better than a gang rape [curra].

For Roberto and Raimundo, women on the street at 3:00 A.M. (cf. Carmen above) are fair game for a maratona. The story focuses on the pleasurable aspects of rape, but in the background is the suggestion "they asked for it." Rubem, 18, an unskilled casual laborer, makes this suggestion explicit. He first describes a maratona in which he participated, then recounts a second that he heard about from friends. In the first case, Rubem is walking on the beach, a well-known trysting spot, with his *turma*—his neighborhood age-set. It is late at night. The men stop a couple at knifepoint and announce their intention to rape the woman, first trying to strike a deal with her companion.

> They said that they wanted the woman and so on, that the guy had to cooperate. He would go first, but then they'd follow. The guy said no, he said he'd give them 5,000 [cruzeiros]; 5,000 at that time [the early 1980s] was an awful lot. So then Lauro said he wouldn't accept the 5,000, the others would, but not Lauro, he said that what he wanted to do was fuck [laughs]. What he wanted most was to satisfy his desire, even by force.

There were about five or seven of them. They'd already agreed not to use each other's real names. They went one at a time. We stood guard, to see if anyone came. Everyone was quiet. The girl didn't say anything [either], she was crying. She didn't like it, you could see she didn't like it.

[This girl] was well-behaved [*comportada*]; [this maratona] made the radio. Maratonas are totally illegal, to take a person by force. [But with some girls, you'll never have a problem with the police.] The easiest type are those that use marijuana, because there are certain girls who are 12, 14, 15, or older; sometimes they get stoned and then surrender themselves. But there are others, the real druggies, who'll do it without getting stoned. You already know about them, and you attack them, even by force, even without them wanting it, there's no problem [laughs].

[There was another time,] there was Mamão, Bocado, Cobra [nicknames of friends], maybe some others. They already knew about her thing, that she'd do it for smoke [marijuana]. I mean she doesn't value herself, I think that's how they all thought, so they went and grabbed her. When they took her, it was against her will; but after she got there, she felt right at home [laughs].

The woman on the beach is someone "well-behaved"—she is not a druggie or a prostitute, although there is a presumption of sexual looseness in her very presence on the beach at that hour. She "didn't like it"; she never "felt right at home." This was a dangerous act for the men: Because she was a good girl, the crime made the radio and could have drawn the serious attention of police. The woman at the party, on the other hand, did not "value herself": She would do it for marijuana, or, if taken by force, eventually she would enjoy it. I sometimes heard men say that a victim of a maratona "deserves" (*merece*) to be assaulted in this way. They portray a maratona as a punishment for dissolution.

Rape here thus has a strongly punitive cast. The retributive use of gang rape is well-documented cross-culturally: consider, in Brazil for example, Mundurucú and Mehinaku men's assaults on women who view sacred musical instruments or otherwise flout male authority (Murphy and Murphy 1974:100–108, Gregor 1985:100–104). The maratona also has affinities with the *charivari*. In peasant communities of medieval Europe, young men often organized themselves into groups known as Societies of Fools. Despite their self-conscious mockery of authority, such groups sometimes enforced local norms through acts of public ridicule. For example, they would clang pots and pans, shout obscenities, or pile cow manure against the doorway of an older woman who had, in scandalous violation of custom, married a young man (Bailey 1991:84–87).

Compared to the maratona, the malicious antics of the Fools seem tame. The chief intent of a maratona is not to create a laughingstock, or even, as in the case of Mundurucú or Mehinaku gang rape, to punish infractions or change behavior. Although past transgressions can provide a useful pretext for a maratona, in the last analysis the victim's actual "guilt" is irrelevant. *For whatever her personal history, the maratona conveniently produces the "crime" of promiscuity in the act of castigation.* More than a punishment, gang rape in São Luís is an act of gender terror. Susan Brownmiller has written that rape "is nothing more or less than a conscious process of intimidation by which *all men* keep *all women* in a state of fear" (1975:5, her emphasis). The allusion to a gender conspiracy seems questionable, but the main point—that male acts of sexual aggression reinforce women's sense of vulnerability—is credible.

In *Fraternity Gang Rape*, Peggy Reeves Sanday suggests that fraternity gang rape is not only a "display of the power of the brotherhood to control and dominate women" but also a form of "male sexual bonding" (1990:10). She emphasizes the homoerotic aspects of gang rape, in particular its substitution of a female surrogate for the real but forbidden objects of sexual desire, the other male participants. "By degrading and extruding the woman who has been the ostensible object of their mutual sex act," she writes, "the men degrade and extrude forbidden sexual feelings from the group" (p. 59). This conjecture must be significantly qualified in the Brazilian context, where males are acceptable, or at least marginally acceptable, sexual objects. In São Luís, I heard of maratonas that had male victims, and such events are of course common in prisons in many countries. What is forbidden as unmasculine is not penetrating another male, but being penetrated. In Brazil, those who penetrate other males do not necessarily consider themselves to be homosexual (Parker 1991:45–47). Rather, the man penetrated is symbolically feminized. One's masculinity depends not on the gender of one's sexual partner but on one's ability to penetrate with the phallus—that is, not on the object but on the act, construed as invasive and humiliating. Under such circumstances, the desire to dominate seems a more convincing motivation than the need to displace erotic feelings toward one's partners.[12]

A maratona, then, seems to be a ritual of power enacted in the idiom of sexuality. Perhaps intoxication with power fuels sexual lust. If these conjectures are correct, a maratona is significantly *about* power, as power in its bad aspect is conceived locally—as invasion, penetration, subjugation, violation, nullification. Conversely, powerlessness is a state of abject vulnerability. In a maratona, meanings of power are violently engraved on the most intimate and sensitive surfaces of the body. As in Franz Kafka's story "In The Penal Colony" (1971 [1919]), the victim's crime—ultimately, the crime of powerlessness—is etched into flesh. If human beings are, in a manner of speaking, inscribable by culture (Kroeber 1917:178), one can hardly imagine a more harrowing technique of cultural inscription.

SEQÜESTROS:
"YOU'LL SEE DAYLIGHT WITH
YOUR MOUTH FULL OF ANTS"[13]

Like the enforcers of gender order, the official police all too often use capricious criminal acts of their own to assert their authority. Abductions and killings by groups of armed men in civilian clothes are, unfortunately, a common occurrence in São Luís as they are elsewhere in Brazil (Chevigny 1990, 1999; Amnesty International 1990, 1994; Human Rights Watch 1994). Local newspapers publish photos of corpses found at the city's desolate margins, blaming the killings on death squads (*esquadrões da morte*) or extermination groups (*grupos de extermínio*). Both press and populace think of such squads as contingents of police engaged in unofficial terror. As I have noted elsewhere (Linger 1990:74; 1992:96, 185–90), police in São Luís have a largely deserved reputation for violence: without question, police even on official business sometimes murder suspected criminals, both on the street and in the jail. I doubt that all killings identified as death-squad murders are the results of police action—police-style executions are also a specialty of the many *pistoleiros* (hired guns), who operate in and around the city, and there may be others who take advantage of the death-squad genre. But in most cases the victim's identity, the *modus operandi,* and the apparent lack of serious investigation cast suspicion directly on the police.

I prefer to term such unofficial police contingents *terror squads.* When terror squads kidnap they do not always aim at killing the victim; more often, their intent is to frighten the person into compliance with the wishes of the authorities, or, simply, as in the following case, to create an atmosphere in which political or criminal deviance is perceived as extremely risky. Toward the end of my mid-1980s fieldwork in São Luís, I approached Juvenal, a man in his 20s, about a kidnapping that I knew he had undergone in 1981. Even five years later, he was apprehensive about the proposed interview, as was I, knowing how the police regarded inquiries into their behavior. Juvenal and I agreed that only pseudonyms would be used and that I would not transcribe the tape until I had returned to the United States.

Juvenal was abducted in 1981 during a period of renewed unrest in São Luís over the half-price student bus fare. The half-fare had been a victory secured after a major rebellion in the city two years earlier. The 1979 riots, in which scores of people were injured (although, luckily, none killed), had thrown the city into bedlam for many days. It was a watershed political event for those who had grown up in the miasma of political repression that had enveloped Brazil since the 1964 coup. Juvenal, like many and perhaps most high school and college students in São Luís, had participated in the 1979 street demonstrations that traumatized the city and state governments but came to symbolize people power among those students and many other city residents.

In late 1981, students once again organized protests against bus company subterfuges aimed at elimination of the half-fare. Juvenal, then a university student in his 20s, was a sympathetic participant in the new demonstrations, but he was not a leader or key player. In early November, he was among those gassed by police in one of the city's main squares; he managed to sneak through the police cordon and was never arrested. Several days later, however, a group of men, almost certainly police, forced him and two friends into cars and took them to a remote beach, where they were beaten. One of the friends, Cláudio, a political activist, was singled out for especially harsh treatment even though, in fact, Cláudio was not at the time a leader of the new movement. Juvenal's experience, which he discusses below, was comparatively mild but nevertheless terrifying:

> [It was] Friday evening, about 6:00. We went over to the *Colégio* [high school] X here in São Luís, to meet one of my friends who was in our [philosophy] study group at the university. There were four of us, and we met [there] most nights, staying until the early hours. At times we even slept there, but by chance that night there were some *padres* [priests] visiting who were staying there, so we headed home. We left at ten, on foot. We dropped off one friend at the bus stop, and kept going. All of a sudden, I saw two cars stopped in the street. Two cars stopped, and some men without shirts, and others with something like a bag on their heads, and suddenly I thought, Hey, there's no soccer game tonight—it was near the municipal soccer stadium. I thought it was strange, at that hour. We crossed to the other side of the street, and they came after us all together, with guns drawn. They grabbed us and forced us into the car. There were eight of them, in two cars. One friend and I went in one car, and my other friend, Cláudio, went in the other.
>
> As soon as we were in the car, they pushed my head down, almost between my legs, I think so I wouldn't see anything. Then they grabbed a cloth, put it around my neck, and tightened it so that I couldn't talk. I soon lost any sense of where we were. The friend who was with me tried to resist; right away they punched him. I [didn't react]; I was afraid of

being hurt. Any movement on my part and they could strangle me. I was thinking about what I could do when we got there, so I wouldn't get beaten too badly.

The police didn't say anything. They didn't want to hear anything we had to say, nothing. We rode a long way, then suddenly I heard the sounds of vegetation beating against the car. I started to get worried. When we stopped, one grabbed me by the arms, behind my back like this, and took me in one direction, and another grabbed my friend and took him the other way. We turned off the asphalted road onto another road that was deserted. They took me there and punched me in the back. Then they said, "Lift your head! Lift your head!" I thought they were going to hit me in the face. I lifted my face; they punched me in the stomach, and that's when I pretended to pass out on the ground. They tore my shirt off, lifted me up, and then let me go, throwing me on the ground. I stayed there as if unconscious. When they left, I raised my head and watched them. When they'd gone some distance, I took off my shoes and ran. I ran more than 50 meters along the road and then into the bush. I stayed there, [hoping] they wouldn't find me. I was full of thorns, in my body, my arms, my legs, I don't know what my face was like. I stayed there about 20 minutes. Then suddenly my friend started calling me: "Juvenal, Juvenal." I managed to find my friend and returned to the road; I got my notebook, my shirt, my shoes. Then we headed through the bush and started walking. We didn't have any idea where we were. I asked him how he was. He'd got a punch in the chest, just a punch. Since he was OK, [we began to think about] Cláudio. We had no idea where he was. The other car had gone on; there was another road, and it had gone off that way. Even if we went to that place, we would run the risk of [getting] beaten up again, or something even worse. So I thought things over and said, "He could even be dead, who knows, but we [have] our own lives [to worry about]." And after all, Cláudio had already been a militant for some time; we had practically no experience of this kind of thing, it took us by surprise, so he'd understand.

At about dawn, we went to the nearest house and asked the man where we were. He showed us where to get the bus. We were still worried about what had happened to our friend. We got the bus, went back to town, and went to my house. Then we went to the colégio. When we got there, Cláudio had arrived. They'd been more violent with him. They'd asked him who were the padres who were involved. He gave [false] names. He gave false addresses of where they were meeting, information that was distorted. So they told him they'd check those places and, if he was lying, they'd come back to kill him. They said, "You'll see daylight with your mouth full of ants!" They beat him a lot. They cut his ears, beat his buttocks until they bled. They took everything from him, leaving him in just his underpants. He walked I don't know how far till he came to a recreational club

on the beach. He convinced the security guards that he'd been the victim
of a kidnapping. Finally, they let him enter. He telephoned his friends, and
they went to pick him up. Then they went to the colégio.

We had breakfast there with the priests. Then right away we went to
the police, to make a complaint. But nothing really happened after this.
Nothing *could* happen. We're almost certain that group was from the police. The methods they used, it couldn't be just anyone, doing the thing
like that. They beat up Cláudio with those nightsticks wrapped in cloth,
so as not to leave marks, all that kind of thing.

Juvenal's traumatic experience, although it took place in 1981, could
have happened yesterday. Acts of abduction, torture, and murder have continued to occur throughout Brazil, long after the formal end of authoritarian rule (Scheper-Hughes 1992: Chapter 6; Chevigny 1990, 1999; Amnesty
International 1990, 1994; Pinheiro 2000; for Argentinian parallels, see
Suarez-Orozco 1992, Timerman 1981). The targets are political activists (especially rural organizers) or perceived human pollutants (suspected criminals, gay men and lesbians, or even street children)—mainly, but not
exclusively, working and nonworking poor people.[14]

The parallels between maratonas and seqüestros are striking. Men kidnap
Juvenal at gunpoint and take him to a stretch of wasteland; men kidnap Carmen at knifepoint and take her to a mangrove swamp. The men punch Juvenal in the stomach and rip off his shirt; the thugs cut Cláudio's ears and beat
his buttocks until they bleed, leaving him in his underpants; Carmen's assailants slash her with a knife, order her to remove her clothes, and rape her
repeatedly. Linguistic evidence, too, reinforces the strong implication that seqüestros allude to sexual aggression. *Porrada,* a blow to the body of the kind
suffered by Juvenal, derives from *porra,* a vulgar term for penis or sperm (Ferreira 1975:1117). A policeman's billy club, used to beat Cláudio, is a *cacete* (or
cassetete), slang for penis. The stereotypical death threat "You'll see daylight
with your mouth full of ants" offers a grotesque metaphor for sexual violation.

Such a threat was carried out in one of the many cases reported by
Human Rights Watch (1994:69–71). On December 28, 1991, Edson
Damião Calixto, a 14-year-old boy from Pernambuco, was accosted by military police who suspected him of a supermarket robbery. The policemen
tore off his jacket and shorts, beat him, handcuffed him, and drove him to
a scrap heap. After threatening to shoot Edson in the mouth if he did not
give them the information they wanted (which he did not have), they forced
him to kneel, and shot him three times. The van drove off, but then returned; one of the policemen shot him again twice, declaring, "Now he's

dead." Shortly before dawn, the van returned once more. The policemen threw the supposedly lifeless Edson into the back and took him to a garbage dump. Here they covered him with trash, thinking that the dump's bulldozer would bury him and his body would never be found. "When Edson realized that his body was covered by ants," the report continues, "he tried to get up, but could not. He also tried to scream, but could not" (1994:70). Miraculously, Edson lived; many others whose terrible experiences appear in the human rights literature did not.

If the seqüestros described by Juvenal and Edson unmistakably draw on the multiple rape scenario, at the same time they mimic the horrifying torture routines widely employed by the 1964–85 military regime. These featured the kidnapping of victims, their removal to unknown locations, their isolation amidst groups of threatening men in ominous and macabre settings, and the administration of electric shocks and physical blows to the sexual organs, often accompanied by violent penetrations of body orifices. Such techniques translated the overwhelming power of the repressive apparatus into bodily experiences of castration and sexual violation, thereby generating a profound sense of vulnerability and degradation.[15] Sexual transgression thereby fused with violent transgression in a searing display of power.[16]

THE END OF THE WORLD

Maratonas and seqüestros assert hierarchical relationships (men over women, armed state agents over dissidents) through threats against and violations of the physical integrity of subalterns and resisters. In this sense, they are acts of political coercion. But they are simultaneously zones of transgression—interactional domains of wild power. Because they dramatize wild power in bodily terms, grounding it in actual or imagined sensations, they lend it immediate emotional force. I am suggesting that such events tend to entrench "the body in the mind" (Johnson 1987), establishing powerful bodily metaphors for the relational abstraction of power.[17]

Hence unlike collective Durkheimian (1965) or Turnerian rituals (1967), which produce solidarity from bodily experience, maratonas and seqüestros use the body to magnify and insist upon social divisions and power inequalities. Although few people actually participate in such events, everyone knows they occur. The imagination feeds on a daily diet of sensational newspaper stories, first-hand accounts, and rumors of police harassment and brutality. In São Luís, as in many other areas of Brazil (cf. Scheper-Hughes 1992: Chapter 6), the narratives are rife with physical detail. Newspaper reports, for example, prominently feature descriptions and even photographs of grotesque bodily injuries and mutilations, inviting the reader's identification with victims or, perhaps for some, with aggressors (Linger 1992). Such accounts not only bring power home, they project it into the private recesses of the body.

In this hierarchical universe, direct confrontation produces victims by penetrating, violating, feminizing, and, sometimes, annihilating. I think this is one reason why, as Roberto DaMatta notes, "common sense tends to define such situations as a *fim do mundo* ('end of the world'), a *Deus-me-livre* ('God forbid'), or as a *Deus-nos-acuda* ('God help us'), that is, as a moment in which the rules of everyday life are suspended and the people involved, often filled with indignation and rage, are left to face themselves and each other" (1991 [1979]:162). DaMatta here focuses on the symbolic obliteration of persons in everyday "authoritarian rituals." Such rituals take the following general form. Someone important (wealthy, highly placed, and well-connected) finds himself[18] in a situation ostensibly governed by universal rules or laws: waiting in a bank queue, submitting to a customs inspection, driving on the highway. The big shot (*medalhão*) ignores the applicable regulations. An interlocutor, thinking the big shot is just an ordinary, anonymous individual, invokes the law against him, whereupon the medalhão thunders, "Do you know who you're talking to?!" He thereby claims an exemption from the rules, underscores his privileged social rank, and asserts his superiority to and authority over the interlocutor, creating a zone of transgression. He moves straight to the teller's window, passes directly through customs, brushes aside the citation as if the interlocutor did not exist. As DaMatta puts it, big shots "shout out the marks of their social identity in order to reduce their adversary to nothingness" (162).[19]

Because such proclamations declare ostensibly universal rules null and void, one should never issue a "Do you know who you're talking to?" from a position of weakness. Euclides da Cunha's chronicle of a nineteenth-century messianic community (1944 [1902]) describes a "Do you know who you're talking to?" gone disastrously awry, eventuating in a local Armageddon. The prophet and his followers declare their opposition to and defiance of the new Brazilian Republic, establishing a theocratic settlement at Canudos, in the backlands of Bahia. The move is a fatal miscalculation. The Republic counters with a "Do you know who you're talking to?" of its own, dispatching column after column of heavily armed soldiers to reestablish the law through a crushing display of wild power. In the brutal, no-holds-barred confrontation, the backlanders, however brave and determined, haven't a chance. Cunha details the army's extermination of the group and the razing of Canudos. His unmistakable empathy for the victims is accompanied, indeed overshadowed, by a sense that what is wrong in the polity runs deeper than the specifics of who wins and who loses such political engagements. The drastic way power is deployed is itself a horrific crime. Despite, or more likely because of, its portrayal of mad destruction, the book is a Brazilian classic—a riveting but disturbing national narrative.

A century has passed since Cunha issued his indictment of wild power; Canudos is a distant memory. The Brazilian state, at present, has no army in the field bent on crushing renegade backlanders, nor is the state, at present, systematically engaged in the torture of dissidents. But violent "end-of-the-world" events are unfortunately common facts of everyday life: decimations of self-respect, beatings of wives, rapes at gunpoint, killings of petty criminals, murders of street children, corpses on the beach, carnage on the streets, so-called suicides in jail. Wild power replicates in these nooks and crannies of Brazilian life. To be sure, there are much more violent nations than Brazil—the world is a nasty place—but for a country at peace, of middling income, and with aspirations to first-world status, Brazil has a disturbingly high incidence of violence, which Brazilians themselves are the first to recognize and deplore.

An example of such Brazilian self-reflection is a song by one of Brazil's most famous, and most incisive, composers of popular music, Chico Buarque. Chico's "Geni and the Zeppelin" (*Geni e o Zeppelim*) provides a devastating critique of wild power from within, a critique consonant with my own but couched in sardonic, surreal Brazilian cultural imagery.

"GENI AND THE ZEPPELIN"

Popular music, like all popular art forms, draws upon cultural common sense. Successful popular music does so in striking ways; it distills, reveals, and often criticizes as it entertains. "Geni and the Zeppelin" first appeared in Chico's play *The Scoundrel's Opera* (*Opera do Malandro,* 1978). The drama takes place in Rio de Janeiro during World War II, when the dictator Getúlio Vargas ruled Brazil. Although discreetly set in the past, *The Scoundrel's Opera,* like much of the author's work of the 1970s to mid-1980s, is a denunciation of the military dictatorship then governing the country. But Buarque's creations can never be reduced to purely political commentary. For him, politics is always bound up with sexuality and the absurdity of the human condition.

The Scoundrel's Opera is an imaginative, caustic, playfully serious Brazilian version of *Three-Penny Opera,* populated by colorful characters and permeated with a Brechtian political sensibility. It unfolds in an atmosphere of thoroughgoing political repression and corruption. "Geni and the Zeppelin" is sung by Geni himself, a character in the play, accompanied in the refrains by a chorus. The song is a fantasy about a persecuted hero(ine) likewise named Geni. By implication, the Geni of the song is, like the Geni of the play, a *bicha,* a "passive" homosexual. Throughout the song Geni is referred to uniformly as "she." We can infer that Geni is ambiguously woman/bicha.

Buarque models "Geni and the Zeppelin" after Brecht and Weill's "Pirate-Jenny," sung by the prostitute Jenny in *Three-Penny Opera* (Columbia Records n.d.). "Pirate-Jenny" is a dream of salvation. A mysterious ship with eight sails and fifty cannons arrives unannounced in the harbor, bombarding the town but sparing Jenny's hotel. The crew spills into the streets, rounds up the survivors, and brings them before Jenny in chains. Asked whom to kill, she orders: "All of 'em!" and celebrates the rolling of heads before sailing off with the brigands. In sum, the song narrates Jenny's apocalyptic fantasies of vengeance and flight.

Chico's "Geni and the Zeppelin" is bleaker. Like her counterpart, Geni suffers indignities at the hands of the townspeople. But when the death ship, a frigid specter of power, appears overhead, things only get worse. The zeppelin's commander demands a night with Geni as the price for sparing the town. She finds him repellent, but reluctantly gives in to the citizens' tearful supplications. She submits to a disgusting nightlong sexual marathon with the commander. Having satisfied his crude desires, he vanishes at daybreak. Geni's sense of relief is shattered when, a moment later, the people resume their persecution, hurling vilifications and stones at her. No privateer of deliverance is going to sail into Geni's Brazilian harbor.

The song introduces Geni as the lover of all those on the "black and twisted" margins of city life: the blind, migrants from the backlands, lesbians, pederasts, crazies, prisoners, lepers, sick old men, and forlorn widows. She is a "well of goodness"–and, precisely for this reason, the citizens ceaselessly torment her:

> Throw stones at Geni
> Throw shit at Geni
> She's made to beat up on
> She's good to spit on
> She puts out for anyone
> Goddamned Geni!

One day an enormous, shining zeppelin appears in the sky. As it hovers menacingly over the buildings, two thousand orifices open in its skin, from which two thousand cannons protrude. The inhabitants are paralyzed with fear. The zeppelin's commander announces: "When I saw all the horror and iniquity in this city, I decided to blow it up. But I'll relent, if that beautiful woman agrees to serve me tonight."

The woman, of course, is Geni. To the astonishment of the citizens, she has unwittingly captivated the imposing chieftain. But there's a problem. Our innocent "virgin," Chico tells us, has her own secret caprices. She would rather make love to animals than sleep with this copper-scented warrior. The citizens, appalled at her "heresy," make a pilgrimage to Geni, pleading with her as they kiss her hand. The mayor kneels; the bishop weeps; the banker offers her a million:

> Go with him, go, Geni
> Go with him, go, Geni
> You can save us
> You'll redeem us
> You put out for anyone
> Blessed Geni!

Faced with such sincere, heartfelt pleas, Geni overcomes her disgust and surrenders, "as someone who gives herself to the hangman." The captain subjects her to a night of "filth" (*sujeira*), until he is finally satiated. Then, at the break of dawn, the silver zeppelin flies off into a "cold cloud." Geni gives a sigh of relief, rolls over, and smiles weakly—but the city will not let her sleep:

> Throw stones at Geni
> Throw shit at Geni
> She was made to beat up on
> She's good to spit at
> She puts out for anyone
> Goddamned Geni!

Geni's story conflates the rape and abduction scenarios. A "well of goodness," Geni inhabits the social depths: above her rank the citizens who torment her, then the city's aristocracy (the mayor, the banker, the bishop), and finally the zeppelin's commander. Masculinity, violence, and criminal privilege order this repressive hierarchy. The zeppelin hovering overhead, with its thousands of cannons jutting from thousands of orifices, could hardly be more dominant, coercive, or phallic. A parallel is drawn: The zeppelin threatens the city with annihilation; the commander violates Geni.

In short, the commander is a rapist/cop/chieftain—the ultimate big shot. Like the Republic's army of liquidation, he proposes to extinguish the "horror" of the city by raining down horror upon it. Armed to the teeth in his splendid airship, he ultimately substitutes rape for slaughter, one form of power-lust for another. His metal vehicle, his lethal weapons, and his shining armor bring to mind the military (the supreme rulers), the police (their agents), or other males with access to powerful machines and powerful arms (assailants such as Roberto and Raimundo, or those who forced Juvenal into their car). Terror, destruction, and phallic aggression are conjoined in a single overwhelming, violent figure who bends others—men (mayor, banker, bishop) or women (the ambiguous Geni)—to his wishes.

Chico treats the other powerful figures in the story—the mayor, the bishop, the banker—with derision. These craven hypocrites get a brief comeuppance when they are symbolically feminized by the commander. But Chico's song is neither a lament for the victimization of the masses nor a romantic celebration of their resistance: The moment the zeppelin disappears the castigation of Geni resumes. Those terrorized by the commander in turn terrorize Geni, not with the gleaming cannons of the powerful but with the rude weapons of the weak: insults, stones, spit, blows, shit—"gratuitous acts and extravagant defilements."[20] In their own petty way, the citizens, like the commander and the perpetrators of maratonas in São Luís, are simultaneously rapists and cops.

There is one hero(ine) in this story: Geni. But her heroism is equivocal, for Geni, refusing to be either rapist or cop, begins and ends the story as a victim.[21] Buarque's dismal tale denounces the coercive actions of the commander, mayor, banker, and bishop, but he targets also the proliferation of wild power that makes victims *and victimizers* of even the oppressed. Chico offers no answers; Max Weber's famous webs of significance appear here more ominously as webs of power, enmeshing everyone in a suffocating authoritarian logic.[22]

POSTSCRIPT: WONDER WOMAN

Encouraging changes have occurred in São Luís since my first field trip in 1984–1986. Local women's groups, for example, were instrumental in founding the DEM, the special police station dealing with crimes against women, in 1988. The DEM has certainly made it easier for women to bring actions against men, and men I talked to in São Luís were well aware that being subpoenaed by the DEM meant trouble. Yet . . . I am uneasily reminded of a new city legend. During my 1991 visit, some men spoke to me jestingly, but uneasily, of one of the women police officers of the DEM. This officer, nicknamed "*Mulher Maravilha*" (Wonder Woman, the cartoon character), was said to kill men coldly and capriciously. No man wanted to come face-to-face with Wonder Woman.

Wonder Woman had become, in the popular imagination, a frightening avenger, a semi-mythical figure who resembled both the legendary bandits of the northeastern backlands and the legendary torturers of the jailhouse basements. Wonder Woman, like Pirate-Jenny, was a juggernaut. It is tempting to celebrate figures such as Wonder Woman, for the victims "deserve it." But in the legend of Wonder Woman, we see an acute dilemma of resistance. In remaking relations of power, how does one also remake practices and conceptions of power?

ACKNOWLEDGMENTS

F. G. Bailey, Carolyn Martin Shaw, Ruben Oliven, and Mark Cravalho provided helpful criticisms of this essay, a version of which I presented at a University of California, San Diego anthropology colloquium in April 1994. Carlos Ramos interviewed Jair and Rubem. Thanks to the Delegacia Especial da Mulher in São Luís (especially to Comissária Además Galvão de Carvalho Lima) and to members of the city's Grupo de Mulheres. Grants from the University of California, Santa Cruz and a Fulbright-Hays Dissertation Research Fellowship supported the research.

NOTES

1. The estimated population of the county (*município*) of São Luís was 781,374 as of July 1, 1991 (IBGE 1991).
2. Zones of transgression were not, however, inventions of the Brazilian military. A thread of sociocultural continuity stretches from the colonial sugar plantation, whose slavocrat master was "virtually a temporal god" who ruled by "fiat" (Freyre 1964:161), to the torture center, whose masters were self-declared temporal devils similarly unbound by bureaucratic rules and restrictions.
3. I do not mean to suggest that wild power is supreme or uncontested (see, for example, the song analyzed at the chapter's end). But it is tenacious; dislodging it from micropolitical settings requires, I think, a more thoroughgoing shift than a change in regime.
4. The first women's police station in Brazil was founded in São Paulo in 1985. See Hautzinger (1997) for an ethnographic discussion of the women's police station in Salvador, Bahia.
5. São Luís DEM statistics show 108 reports of rape in 1990 and 101 of attempted rape, but these figures are surely understated. No separate statistics are kept on gang rapes.

6. In a 1990 case absent from the police reports, but which surfaced in the newspapers, four unemployed young men accosted a 26-year-old woman, beat her until she passed out, raped her, burned her face with cigarettes, inserted pieces of tile into her vagina, and bit off one of her nipples (*O Imparcial,* June 7, 1990).

7. See Parker (1991) and Linger (1992:141–42).

8. Carmen's treatment, brutal as it was, may have been mitigated by her appeal for "clemency" on the grounds that she was a mother of three children—a valued female status opposed to that of *puta* or *vagabunda.*

9. Nevertheless, in almost all cases I examined the sexual aggression was intraclass—both the victims and the victimizers were poor.

10. According to Americas Watch (1991:45), only two men were convicted and punished for crimes of violence against women between 1988 and 1990 in São Luís—out of 4,000 complaints filed with the women's police precinct!

11. Depositions of accused men tend to be unenlightening: usually they claim not to have touched the woman, attributing the crime to the others and sometimes positioning themselves as unwilling participants or even protectors.

12. Brownmiller (1975:296) argues that prison rape originates in "the need of some men to prove their mastery through physical and sexual assault, and to establish . . . a coercive hierarchy of the strong on top of the weak."

13. "*Você vai amanhecer com a boca cheia de formigas.*"

14. Reports of violence against gay men and lesbians include Amnesty International (1993b), Grupo Gay da Bahia (1993), and Mott (1987, 1988a, 1988b). Dimenstein (1991) discusses the plight of street children.

15. Both men and women were customarily subjected to such techniques. See Arquidiocese de São Paulo (1985) for documentary evidence; the book, although matter-of-fact in tone, provides appalling details. Gregory and Timerman (1986) discuss ritual aspects of torture in Argentina during the "dirty war."

16. Parker (1991) addresses the intersections between sexual transgression, violence, and power.

17. Cf. Turner's (1967) idea of the sensory and ideological poles of symbols and Sapir's earlier notion of "condensation symbolism," which, says Sapir, "strikes deeper and deeper roots in the unconscious, and diffuses its emotional quality to types of behavior and situations apparently far removed from the original meaning of the symbol" (1934:494; passage quoted in Turner 1967:29).

18. The prototypic "big shot" is a man, although women can also under some circumstances play this game.

19. Depending on the situation, almost anyone can be a big shot. In street and bar fights (*brigas*), the "Do you know who you're talking to?" often takes the

form of a physical assertion made with fists or knives (Linger 1992). DaMatta suggests that ordinary people sometimes proclaim "Do you know who you're talking to?" through protest riots, destroying property and disobeying laws en masse (1982:41).

20. Brownmiller (1975:213), quoting Woods (1969).

21. Geni is not exactly powerless—she does, after all, save the city—but the power she exercises entails her own martyrdom. There is a suggestion here of a minor concept of power, with overtones of sacrifice, perhaps associated more strongly with femininity. This topic merits further exploration.

22. See Clifford Geertz's (1973:5) memorable characterization of Weber's approach to meaning.

BIBLIOGRAPHY

Abers, Rebecca
1996   From Ideas to Practice: The Partido dos Trabahadores and Participatory Governance in Brazil. Latin American Perspectives 23(4): 35–53.
Alvarez, Sonia E.
1990   Engendering Democracy. Princeton: Princeton University Press.
Alvarez, Sonia E. and Arturo Escobar, eds.
1992   The Making of Social Movements in Latin America. Boulder, CO: Westview Press.
Americas Watch
1987   Police Abuse in Brazil: Summary Executions and Torture in São Paulo and Rio de Janeiro. New York: Americas Watch Committee.
1991   Criminal Injustice: Violence against Women in Brazil. New York: Americas Watch Committee.
Amnesty International
1990   Brazil: Torture and Extrajudicial Execution in Urban Brazil. New York: Amnesty International.
1993a "Death Has Arrived": Prison Massacre at the Casa de Detenção, São Paulo. New York: Amnesty International.
1993b Gay Human Rights Newsclips: Brazil. AIMLGC Newsletter 3(9): 7. Santa Cruz, CA: Amnesty International Members for Lesbian and Gay Concerns.
1994   Beyond Despair: An Agenda for Human Rights in Brazil. New York: Amnesty International.
Andrews, George Reid
1991   Blacks and Whites in São Paulo, Brazil 1888–1988. Madison: University of Wisconsin Press.
Ardaillon, Danielle and Guita Grin Debert
1987   Quando a Vítima é Mulher. Brasília: Conselho Nacional dos Direitos da Mulher.

Arquidiocese de São Paulo.
1985   Brasil: Nunca Mais (2nd ed.). Petrópolis: Vozes.
Bailey, F. G.
1991   The Prevalence of Deceit. Ithaca, NY: Cornell University Press.
Brownmiller, Susan
1975   Against Our Will: Men, Women, and Rape. New York: Bantam Books.
Buarque de Holanda, Chico
1978   Opera do Malandro. Lisbon: Edições "O Jornal."
Chevigny, Paul G.
1990   Police Deadly Force as Social Control: Jamaica, Argentina, and Brazil. Criminal Law Forum 1(3): 389–425.
1995   Edge of the Knife: Police Violence in the Americas. New York: New Press.
1999   Defining the Role of the Police in Latin America. In The (Un)Rule of Law and the Underprivileged in Latin America. Juan E. Méndez, Guillermo O'-Donnell, and Paulo Sérgio Pinheiro, eds. Pp. 49–70. Notre Dame: University of Notre Dame Press.
Columbia Records
n.d   Die Dreigroschenoper. Music by Kurt Weill. Lyrics by Bertolt Brecht. Guy Stern, trans. Columbia Masterworks LP O2L 257.
Cunha, Euclides da
1944 (1902)   Rebellion in the Backlands. Samuel Putnam, trans. Chicago: University of Chicago Press.
DaMatta, Roberto
1982   As Raízes da Violência no Brasil: Reflexões de um Antropólogo Social. In Violência brasileira. Roberto DaMatta, Maria Célia Pinheiro Machado Paoli, Paulo Sérgio Pinheiro, and Maria Victoria Benevides. Pp. 11–44. São Paulo: Brasiliense.
1991 (1979)   "Do you know who you're talking to?!" The Distinction Between Individual and Person in Brazil. In Carnivals, Rogues, and Heroes: An Interpretation of the Brazilian Dilemma. John Drury, trans. Pp. 137–197. Notre Dame: University of Notre Dame Press.
Dimenstein, Gilberto
1991   Brazil: War on Children. London: Latin American Bureau.
Durkheim, Emile
1965 (1915)   The Elementary Forms of the Religious Life. New York: Free Press.
Ferreira, Aurélio Buarque de Holanda
1975   Novo Dicionário da Língua Portuguesa. Rio: Nova Fronteira.
Freyre, Gilberto
1964   The Patriarchal Basis of Brazilian Society. In Politics of Change in Latin America. Joseph Maier and Richard W. Weatherhead, eds. Pp. 155–173. New York: Frederick A. Praeger.
Geertz, Clifford
1973   Thick Description: Towards an Interpretive Theory of Culture. In Clifford Geertz, The Interpretation of Cultures. Pp. 3–30. New York: Basic Books.

Gramsci, Antonio
1971   Selections from the Prison Notebooks of Antonio Gramsci. Quintin Hoare and
       Geoffrey Nowell Smith, eds. and trans. New York: International Publishers.
Gregor, Thomas
1985   Anxious Pleasures: The Sexual Lives of an Amazonian People. Chicago: Uni-
       versity of Chicago Press.
Gregory, Steven and Daniel Timerman
1986   Rituals of the Modern State: The Case of Torture in Argentina. Dialectical
       Anthropology 11(1): 63–72.
Grupo Gay da Bahia
1993   Boletim do Grupo Gay da Bahia 26. Salvador, Bahia: Grupo Gay da Bahia.
Hautzinger, Sarah
1997   "Calling a State a State": Feminist Politics and the Policing of Violence
       against Women in Brazil. Feminist Issues 15: 3–30.
Human Rights Watch
1994   Final Justice: Police and Death Squad Homicides of Adolescents in Brazil.
       New York: Human Rights Watch.
O Imparcial (São Luís)
1990   Desocupados Estupram Jovem e lhe Arrancam o Bico do Seio. June 7, p. 7.
IBGE (Instituto Brasileiro de Geografia e Estatística)
1991   [No title. Mimeographed sheet with population estimates for Maranhão.]
       São Luís: IBGE office.
Johnson, Mark
1987   The Body in the Mind: The Bodily Basis of Reasoning, Imagination, and
       Reason. Chicago: University of Chicago Press.
Kafka, Franz
1971 (1919)   In the Penal Colony. In Franz Kafka: The Complete Stories. Nahum
       N. Glatzer, ed. Pp. 140–167. New York: Schocken Books.
Keck, Margaret E.
1992   The Workers' Party and Democratization in Brazil. New Haven: Yale Uni-
       versity Press.
Kroeber, A. L.
1917   The Superorganic. American Anthropologist 19(2): 163–213.
Linger, Daniel T.
1990   Essential Outlines of Crime and Madness: Man-fights in São Luís. Cultural
       Anthropology 5(1): 62–77.
1992   Dangerous Encounters: Meanings of Violence in a Brazilian City. Stanford:
       Stanford University Press.
1993   The Hegemony of Discontent. American Ethnologist 20(1): 3–24.
Mott, Luiz
1987   O Lesbianismo no Brasil. Porto Alegre: Mercado Aberto.
1988a Escravidão, Homossexualidade, e Demonologia. São Paulo: Ícone.
1988b O Sexo Proibido: Virgens, Gays e Escravos nas Garras da Inquisição. Camp-
       inas: Papirus.

Murphy, Yolanda and Robert F. Murphy
1974   Women of the Forest. New York: Columbia University Press.
Parker, Richard G.
1991   Bodies, Pleasures, and Passions: Sexual Culture in Contemporary Brazil. Boston: Beacon Press.
Pinheiro, Paulo Sérgio,
1983   Crime, Violência e Poder. São Paulo: Brasiliense.
2000   Democratic Governance, Violence, and the (Un)rule of Law. Daedalus 129(2): 119–143.
Sader, Emir and Ken Silverstein
1991   Without Fear of Being Happy: Lula, the Workers' Party, and Brazil. New York: Verso.
Sanday, Peggy Reeves
1990   Fraternity Gang Rape: Sex, Brotherhood, and Privilege on Campus. New York: New York University Press.
Sapir, Edward
1934   Symbolism. In Encyclopedia of the Social Sciences 14: 492–495.
Scheper-Hughes, Nancy
1992   Death Without Weeping: The Violence of Everyday Life in Northeast Brazil. Berkeley: University of California Press.
Suarez-Orozco, Marcelo
1992   A Grammar of Terror: Psychocultural Responses to State Terrorism in Dirty War and Post- Dirty War Argentina. In The Paths to Domination, Resistance, and Terror. Carolyn Nordstrom and Joann Martin, eds. Pp. 219–259. Berkeley: University of California Press.
Timerman, Jacobo
1981   Prisoner without a Name, Cell without a Number. New York: Alfred A. Knopf.
Turner, Victor
1967   The Forest of Symbols: Aspects of Ndembu Ritual. Ithaca: Cornell University Press.
Woods, G. D.
1969   Some Aspects of Pack Rape in Sydney. Australian and New Zealand Journal of Criminology 2(2): 105–119.

CHAPTER 5

RECOGNITION OF STATE AUTHORITY AS A COST OF INVOLVEMENT IN MOROCCAN BORDER CRIME

David A. McMurray

WHILE LIVING IN NORTHERN MOROCCO IN 1986–1987, I often took the occasion to walk across the frontier between the Berber-speaking, duty-free, smugglers' paradise of Nador, Morocco, and its sister city, the old, Spanish-colonial North African duty-free enclave of Melilla.[1] On one such occasion, I came upon a poorly dressed, small-time smuggler. He was loaded down with black plastic bags—full, no doubt, with tea, canned milk, cheese, coffee, whiskey, baby food, chocolate, cookies, canned fruit juices, rice, or possibly with bathroom tiles, shampoo, cologne, shoes, sheets, or maybe even light bulbs. It is difficult to name an item that is not smuggled across that border into Morocco. In any case, I knew he was poor from the condition of his clothes, but also because all of the wealthier smugglers did not have to run the risk of passing through the secured border on foot.

The smuggler I was surreptitiously watching had already made it past the Spanish border guards and was now confronting the Moroccan army checkpoint on the other side of the small border bridge. Here a platoon of soldiers lounged around and harassed smugglers and other poor border crossers. They would relieve petty smugglers of some of their goods on occasion, and were especially harsh on those caught trying to cross the border anywhere but on the road. My man at this point was trying simultaneously to clutch his bags and shelter his face with his hands, all the while crying out for mercy as he got slapped around by a snarling group of the army border guards. I hunched my shoulders, walked straight ahead, and pretended not to notice the alarming and discomforting displays of dominance and subordination taking place all around me.

From the army checkpoint on, I and all other traffic entered a no-man's-land of approximately 100 yards that were decorated, as were many roadways in the country, with some kind of symbol of the regime—in this case, evenly spaced Moroccan flags. On one side of the road stood a flour mill; on the other was a welcome tent set up by one of the state banks. At the end of this section of the border stood the Moroccan passport and customs control building. Border crossers at this point were first stopped in the street by uniformed customs guards, who either directed them to a window to get searched and then have their passport stamped or sent them to a walkway on the side of the road meant for local residents with nothing to declare. It was here that I encountered a bunched-up group of poor women smugglers who were beseeching a bored-looking border guard to let them pass onto the walkway.

The small-time female smugglers' preparations to cross had actually started several blocks away from the border in Melilla in a stand of scrawny pines across the street from a strip of wholesalers' shops. Here, the smugglers would prepare their loads. These poorer female smugglers had found that they had an easier time crossing than men. They used the cultural discomfort of the authorities at having to physically confront a woman to their own advantage by stopping among the pines to wrap the goods to be smuggled on their bodies. In this way, they were assured of at least getting those items across. They knew that the guards would not physically search a woman. When they were finished wrapping, the women often looked enormous— like astronauts, but without helmets. Underneath their *jalabas* hung dozens of tablecloths, towels, linens, bloomers, and sweat suits draped over rope and then tied to their chests. They could even carry glassware and teapots stuffed inside oversized stockings and pants, which they put on layer after layer. The women would put what their bodies could not carry in black plastic bags and haul it by hand, always hopeful that the sexual code of conduct of the male guards could be turned to the point where they would not want to get involved in an unseemly fight with a woman for the contents of her bags.

Normally, the ruse worked. I never saw a guard physically touch one of the female smugglers. They would harangue them and sometimes force them to wait for long hours in the no-man's-land, but they never touched them. For the petty male smugglers, however, it was a different story. The guards would, on occasion, mercilessly beat them. Their trip through the no-man's-land was always an ordeal. Their only trick or "tactic" was the age-old weapon of the poor when confronted by a bureaucrat: obsequiousness. Their profuse fawning, begging, and pleading would normally be rewarded by eventual passage, although they had to endure the pilfering of the guards and the occasional slap and kick along the way.

I was trying to watch these various smuggling operations without myself being observed. I failed. A cop I recognized from town who had been rotated out to the border came up, stopped me, and cheerfully invited me in for a glass of tea in the welcome tent set up by the bank not ten yards from the scenes of public humiliation. Positively oblivious to what was taking place around us, he scolded me good-naturedly for not greeting him as I passed. I can only assume that he had grown so accustomed to the border theater, with its petty but demeaning stylistics of power, that it did not strike him as worthy of any notice at all.

I think I can honestly say that a good portion of the population of Nador viewed not the smugglers but the various state agents as the true criminals. To them, the border guards and cops and army men employed to interdict smuggling were the truly wicked. They were the ones who preyed on the weak and took pleasure in abusing their powers.[2] The fact that they had their price—indeed, as we have seen, demanded their price—meant that they were regarded by the Nadori population as parasitically living off of the smugglers. They robbed the residents of Nador.

I do not want to leave Nadoris with the last word on this subject, however. I think there are other issues not completely teased out by their commonsense take on the border. I want to think, in the rest of this chapter, about borders and border runners (that is, smugglers). I want to try to begin at least to make sense out of the border scene just described to you. I find it interesting that the mere presence of the border between Nador, Morocco, and the Spanish enclave of Melilla creates lawlessness. This state of affairs is, on the surface, curious because the border is the quintessential material form of the state and its law. Folklore, of course, has always associated lawlessness with the space of the border. But, thinking about it logically, one would not expect to find criminality so flagrantly tied to places such as borders which are, after all, locales established for the express purpose of exhibiting state power. On top of this more generalized puzzling relationship between frontiers and criminality there is the specific one of why the Moroccan border guards incongruously beat up and harass petty smugglers but do not confiscate all their goods or completely prohibit their movement.

This is the seeming paradox that I want to try to make sense of here.

Before trying to explain the dynamics of today's relationship between Nador border criminals and the Moroccan state, let me turn back and provide some of the historical context and then a bit of discussion concerning smuggling routes. I want to do so because I think a case could almost be made that historical investigation of relations between the Moroccan state and its Nador hinterland alone goes a long way toward explaining the contemporary prominence of smuggling. The historical record certainly suggests that the region's economy has for too long depended on the involvement of too many people in illegal activities for these to be easily shut down. Yet I also think that theoretical explanations do shed light on the nature of relations in general between border populations and the state—relations of which the Nador case is just one specific instance. So, in the name of broadening the lessons to be learned from Nador, I want to follow the sections on the history and circuits of smuggling with a discussion of what I find theoretically provocative. I will do so by discussing the applicability of three theoretical approaches in the literature on domination and opposition, in general, and on states and borders in particular. I will start with an all-too-brief mention of the model of state control and popular resistance associated with the work of Michel de Certeau (1988), which I think gets it wrong—or at least does not get to the heart of the matter. Then I want to describe in a bit more detail what might be called two variants of a Foucauldian approach. The first, as represented by Lisa Wedeen (1999), takes up the analysis of how an authoritative regime can produce political power and create compliance by means of rhetoric and symbolism. Wedeen's study of the cult of personality in Syria is very instructive, but does not quite fit the case of border crossers in Nador, Morocco, for reasons to be explained shortly. Finally, I will examine the approach of the Cameroonian political scientist Achille Mbembe (1992) and discuss how I believe he comes closest to capturing what I take to be the more interdependent and complex relationship binding borders and smugglers.

HISTORY OF SMUGGLING

Spanish "claims" to the north of Morocco date back to the 1492 Reconquista when excess enthusiasm spilled over from the fall of Granada and caused the Spanish to capture the enclave of Melilla, 12 kilometers from Nador. In 1912, the French formally invaded Morocco and established a protectorate in the south while the Spanish took the north (the lion's share of which was comprised of the mountainous, Berber-speaking region known as the "Rif"). The country of Morocco was liberated and reunited again in 1956—all except the presidios, or enclaves, of Melilla and Ceuta and several small islands on the Mediterranean coast that remain under Spanish control. When the protectorate was handed back to Morocco, the Spanish saw no reason to include the small areas that they had conquered 400 years before.

The presence of this Spanish territory (Melilla) on the northeastern Moroccan coast represents the most important factor behind Nador's contemporary status as an unofficial free trade zone and smugglers' paradise. Of course, smuggling has long been a part of life in the northern, or Rif, region of Morocco. It was originally fueled by the inability of the *Makhzen* (central Moroccan state administration) to control trade along the Rif coast; by European connivance in that illegal trade; and by the proximity of Spain and Gibraltar to the north and the French colony of Algeria to the east. Goods and services have passed back and forth through this rear window between Africa and Europe for most of the twentieth century. Melilla installed customs and was granted the right by the Makhzen to develop commercial relations with the Rif as early as 1866. Boosters of Spanish expansionism in the Rif complained about the lack of profitable trade between Melilla and its recalcitrant Rifi neighbors around the turn of the twentieth century, although, as a 1913 report put it, "there is not a single village or hamlet (*char*), regardless of how far removed, where products sent to their doors by the [Spanish] Empire do not arrive" (cited in Ouariachi 1980:188). Jewish caravans coming up from the interior were reported to be stopping in Melilla once a week as late as 1912 (Slouschz 1927:384). Jewish merchants in the Rif were also important in peddling English goods to the locals, although many Muslim Rifis made the easy trip themselves across the Mediterranean to Gibraltar in order to bring back English products (Ouariachi 1980:189). Older Nadoris, when I was there, still spoke of the brisk trade in food stuffs that used to link Melilla to the country during the Spanish Protectorate.

In any case, in the years following Independence in 1956, the border between Melilla and its hinterland began to close. A Moroccan identity card became mandatory for crossing.[3] At the same time, Moroccan and Spanish authorities started interfering more in the flow of goods. Moreover, the reunification of the northern (ex-Spanish) and southern (ex-French) zones within Morocco proper exacerbated the hardship for the population of the north: the disappearance of tariffs and price differences between the two Moroccan zones, plus the closing of the Algerian frontier, led to a sharp decline in labor migration and smuggling. Tensions increased to the point where a rebellion broke out in the northern provinces of Al Hoceima and Nador in 1958–1959. (Then) Crown Prince (now King) Hassan himself led the counterattack with 20,000 troops backed up by planes and possibly napalm. The repression was severe.[4]

By the late 1960s and early 1970s all of this began to change as Nador's own economy heated up. A decade's worth of repatriated wages from the massive labor outmigration to Europe throughout the 1960s had created the beginnings of a boom. The population doubled between 1960 and 1970

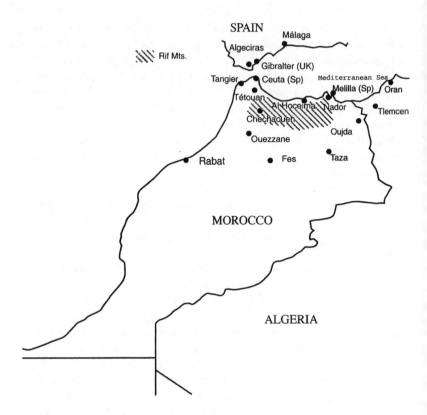

5.1. The frontier between Morocco and Spain.

(from approximately 17,500 to 32,500) from the effects of rural to urban migration, migrant resettlement in Nador, and the growth of the bureaucracy in this provincial capital. The demand for housing fueled a boom in related industries and in the smuggling of foodstuffs, household furnishings, luxury goods—every item imaginable.

The Spanish maintained Melilla's duty-free enclave status throughout this period as an incentive for mainland Spanish tourists to visit. The local population willing to stay in the city and make it their home was obviously also rewarded. The duty-free zone was duplicated informally on the Moroccan side of the border as well. However, the Moroccan state drew the line just outside of Nador. From that point, Moroccan tariffs were enforced and the transport of duty-free goods was illegal. The border between the two cities thus required that merchandise be smuggled across the international frontier from one city to the other and then across the tariff frontier into interior Morocco.

By the latter 1970s the standard of living for many urban Moroccans in the rest of the country had also risen to the point at which they could afford the round trip bus ride to Nador to buy its smuggled goods then try to smuggle them back into interior Morocco themselves. Ouariachi reports (1980:268) that by the mid-1970s over 10,000 people a day were entering Nador from the outlying area and beyond. The population doubled again in the 1970s, the repatriated wages increased considerably, and Nador's construction and real estate speculation-fed economy continued to boom.

By all accounts, smuggling across the international border kept apace. No estimates of the numbers of people who lived by smuggling are available before 1971. In the second half of that year, however, Ouariachi found that 1,358 smugglers were seized (7.5 per day) at the Spanish-Moroccan frontier. Three years later (in 1974), 2,330 smugglers were apprehended at the border during the second half of the year, or 13 per day (1980:246).[5]

By the early 1980s, the Moroccan economy as a whole was once again on the ropes. This time it was suffering from the cost of capital intensive development schemes and the protracted war in the Sahara. A strengthening dollar and falling phosphate export prices added to the country's balance of payments deficit. Debt rescheduling talks became necessary by 1983, during which International Monetary Fund (IMF) authorities suggested that the state reduce its overhead by redistributing misery downwards via price support reductions, as well as by cracking down on contraband in order to help bolster state revenues. The Moroccan government agreed. In August of that year they placed a 500 *dirham* ($60 U.S.) stamp tax on anyone crossing to Ceuta or Melilla as a way of appropriating for the state some of the profits from smuggling. The tax effectively killed the trade.

Prices in Nador immediately began to climb. The following autumn months were tense. Supplies dried up in the *suqs,* and so did the buyers. Skirmishes broke out at the border check point. The frustrated smugglers began to break windows. Customs guards promised that the tax would soon be rescinded, but nothing happened.

The final blow came in the form of cumulative drought. Little rain had fallen since 1979. By the 1983 season, crop yields were catastrophically low in the north. The informal economic activities available to peasants driven into the city of Nador (smuggling and associated transport and peddling) had been destroyed by the border tax.

Lycée (high school) students eventually provided the initiative and leadership needed to organize the discontent. The festering problems of credential inflation and limited professional employment opportunities further increased their frustrations.[6] Rumor had it that school fees were to be raised, so students from around the country (but particularly in Marrakech and throughout the north) began to hit the streets during the early weeks of 1984.

January 17, 1984, began in Nador with another student demonstration headed, as usual, by the students of Lycée *Al-Maatar* (the airport district), the poorest equipped of the three lycées and the farthest from the center city (lack of good transportation to and from campus being a chronic complaint of the students). During the day, thousands of poor and unemployed males tied to the smuggling trade joined the students in the streets. Over the next two days, their numbers swelled to over 12,000. They overturned cars, pulled down telephone lines, and smashed the windows of government offices and shops. Particularly hard hit were the banks, pharmacies, and Westernized boutiques, such as hair salons and Western dress shops. The police, badly outnumbered, eventually barricaded themselves in the downtown police station. The crowd chanted "Down with Hassan II! Long live Abdelkrim! Mohammed V was our father but you, who are you? Enough of prisons and palaces!" (Clément 1986:30).

Reinforcements were slow to arrive because of the security priority represented by the Islamic Conference being held in Casablanca at the same time. When they did arrive some time later, repression was swift. Shoot to kill orders went out, and no adult male was safe on the streets. Interrogators armed with lists of alleged ringleaders began bashing down doors and hauling away students and unemployed smugglers alike. Anyone unable to find shelter was shot or rounded up. Some hardcore demonstrators continued to battle with stones in the narrow alleys of Poblao, the poor section of town up on the hillside. Helicopter gunships eventually flushed them out. The final body count ranged from the official dozen or so to the often cited figure of 1,000.[7] Interestingly enough, and in spite of the severity of the state's reaction, the border was soon reopened, the tax was quietly dropped, and smuggling re-established itself as the life blood of the Nador economy. For all intents and purposes, the smugglers had won the day—this in spite of the fact (claimed by Nador's citizens) that the city continues to have the highest per capita number of police and soldiers of any city in Morocco.

SMUGGLING CIRCUITS

There are three major directions in which smuggled goods circulate around Nador. The first kind of smuggling involves bringing goods into Nador from Melilla for sale either to shop merchants, citizens who have privately con-

5.2. Smugglers selling their wares on the streets of Nador.

tracted with the smuggler, or for sale openly on the street by an itinerant seller. The smugglers involved in this trade almost have to be from either Nador City or Melilla because only they can cross the border without paying a tax. All other Moroccans and non-Melilla resident foreigners must pay to leave Morocco via Beni Ensar, the Nador-Melilla crossing point. Locals, on the other hand, may often times be waved right through.

Retailers of clothes and shoes in Nador obtain as much as 70 percent of their inventory from smuggling. Much of the rest comes from Casablanca or from the artisanal production of traditional clothing (*kaftan; babouche; sirwar* [*silwal* in Arabic]; *jalaba,* and so on). Only 20 percent of the customers are local residents (Barjila 1985:48). Smaller items for the home, such as dishes, cooking pots, and utensils come mainly (60 percent) from Morocco, especially Casablanca, with the remainder being smuggled in across the Melilla border. Larger home furnishings such as wardrobes, couches, and armchairs come mainly (70 percent) through legal national circuits, even if imported at Casablanca (although we once saw a man on a bicycle smuggling a chest of drawers into Nador). Electrical appliances such as fans, sewing machines, television sets, VCRs, ovens, and refrigerators, however, are mostly (80 percent) smuggled in via Melilla (Barjila 1985:49–50). Much of the jewelry sold in Nador is fashioned in Fes or Casablanca and bought by locals. The same is true for pharmaceutical drugs, which are made under license from French companies. Auto parts and repair tools are produced in or imported through Casablanca (90 percent). The rest are smuggled or imported in by emigrants, as are most Mercedes cars and trucks. Mercedes importation and smuggling by emigrants is one of their most important income sources during the summer vacation period (Barjila 1985:54). That is why the make is overrepresented in Nador and why Nadoris call their town the "city of Mercedes" (*thamdinth n Mercedes*).

The second type of smuggling involves taking goods out of Nador and into the rest of Morocco and, to a lesser extent, Tunisia and Algeria. This is necessary because Nador itself is not closely policed—anyone can display for sale or purchase an item within Nador. The law is enforced again at the city limits, however; goods smuggled in from Melilla therefore must be smuggled back out again to go elsewhere.

The third category of smuggling involves smuggling out hashish. Since the marijuana used to make the hash is not grown in Nador Province (but next door in Al Hoceima Province and farther west) and little local consumption takes place, this is not a significant part of the smuggling market. A *Commissaire Divisionaire* squad does exist in Nador to maintain surveillance over the hashish trade as it heads through Melilla, but I suspect they are funded by Interpol or some other international agency more concerned than the Moroccans with hashish destinations (one of the Nador *Commissaire* agents used to work in the U.S. Embassy in Rabat). I know, too, that young European tourists will sometimes come through and try their luck smuggling the drug, although most of it seems to make its way into Melilla and then is peddled there or resold and smuggled out again, bound for Europe.[8] The relatively wealthier Spanish tourist traffic that never leaves Melilla to visit Morocco makes Melilla a better hashish market than anywhere in northeastern Morocco. Likewise, the large number of Spanish military personnel stationed in Melilla, especially the rowdy *Legionarios* (Spanish Foreign Legion, or *Tercio*), make good customers.

I was shown a State Department report on the Moroccan drug trade one day in the American Embassy. It claimed that two-thirds of the hash produced in Morocco was consumed locally and most of the rest was smuggled to Europe. The report estimated that only five percent of the total ever made it to the United States, which I take to be good news, suggesting there is no need for a "war on drugs" in the Rif.[9] The smuggling networks for hash also appeared to be decentralized and, interestingly enough, tied in more and more with the networks of migrants who presumably took some back with them to sell in Europe after their Moroccan holidays—or, at least, that is how it appeared in the arrest records from Europe mentioned in the report. I think that some of the rest of the European traffic goes by way of boat. I say this because I knew of one Nadori-Melilla-Europe hash smuggler who lost a shipment that then washed up on the beach and was found by some kids. The Moroccan authorities were angry at this, so they exiled him to Melilla (all of this was informal, of course). Another ex-migrant was known in Nador to use a large commercial fishing boat to cover up his hash smuggling operation.

Many other hash smugglers obviously existed, but they did not mark the city in any way: no one flashed pieces of hash at passersby in the streets of Nador, as they would in Melilla; and no growers' mafia in black Mercedes drove around intimidating outsiders, as they did in some of the growing regions to the mountainous west around Ketama.

BORDER THEORY

I think the examples of Nador smuggling discussed above begin to illustrate the way the state struggles continuously to induce complicity in state rule while its subjects strive continuously to clear and maintain a space free of that rule. Smugglers have a doubly difficult task facing them in that they must struggle to evade or subvert the borders and bureaucracies of more than one state at a time. It is the nature of this struggle that I now want to concentrate on. Let me begin by rehearsing what I find worth keeping in mind about the general nature of the relationship between states and borders.

Nicos Poulantzas (1980:105) argues that the state under capitalism establishes the frontiers of the nation and thus creates national territory. In the process, the state unifies and homogenizes the nation that is inside. The state accomplishes this, according to Poulantzas (1980:113), by monopolizing national traditions while eradicating the "traditions, histories and memories of dominated nations [now within the territory] . . . The state establishes the modern nation by eliminating other national pasts and turning them into variations of its own history." In other words, the modern state demarcates an inside and an outside and, thus, "insiders" and "outsiders." Paraphrasing Poulantzas further, the state thus historicizes a territory and territorializes a history in the process of creating the unified, homogenized nation.

The Border (both in its international and tariff zone forms) stands as the concrete expression of the "inside" and the "outside" of the state, the embodiment of the "territorializing imperative." This does not explain the lawlessness and anarchy historically associated with borders, however. Perhaps it flows from the well-known axiom that the law creates the criminal—needs the criminal—as the policeman needs the thief and the border needs the smuggler and transgressor. They form a symbiotic coupling.

Another interesting and useful critical reflection on the border and its relation to the state occurs in Michael Kearney's (1991:61) discussion of its Foucauldian properties. Kearney reminds us that international borders create migrants. Borders establish the terrain of the haves and have-nots and then, far from serving as barriers to migration, borders actually serve to discipline the migrants. The border represents one of a series of state forms with which migrants come in contact and which instill in them insecurity and fear and a sense of their minority/foreign status—all in the name of reproducing in the migrants discipline and docility and the acceptance of low wages and lower social status.

As with migrants, so also with smugglers: both are creations of borders. There is one difference, though. Migrants do not necessarily set out to break the law, smugglers do. Smugglers, to be successful, must be able to evade the agents monitoring the state's territory and enforcing its laws against improper entry and importation. The nature of their business demands that they avoid the taxes and tariffs and duties and permits and so forth that, if imposed, would remove the incentive to smuggling. Thus the illegality of the operation is vital to it. Were smugglers not able to slip by or get around the state's border controls, they would be out of business. If they were not able successfully to practice illegal entry back across the border, they would have no income at all. Their livelihood depends on putting to good use the price differentials created by the state's import tariffs and enforced by the state's frontiers.

Perhaps it goes without saying that many local Nador residents care not a wit for the technically illegal status of smuggling. Their historical relationship with the Moroccan state has been sufficiently strained to the point where they bristle at having to obey most any directive from the capital. They certainly do not credit the state with having a monopoly on reason or a morally superior authority to determine for locals what is right and wrong. On top of that, the border guards and most other state agents are viewed in the popular imagination as living illegally/immorally/parasitically off the locals in any case. Who are they, then, to determine and enforce what is or is not legal?

CERTEAU AND EVERYDAY RESISTANCE

The French sociologist Michel de Certeau provides the necessary Franco-philosophical rationale for what has proven to be a very popular binary model in which state domination necessarily entails popular resistance. He and his followers in the "resistance" school do not argue that organized, popular movements rise up as a response to state oppression; rather, they want to draw attention to the myriad local, spontaneous, uncoordinated, opportunistic attempts (de Certeau calls them "tactics") on the part of the populace to avoid having to submit to the powers that be (de Certeau 1988:37). In fact, in this model, it sometimes seems that almost all of the practices of everyday life can be construed as somehow "beating the system" or as bluffing, cheating, evading, and eluding official control.

In the case of Melilla and Nador, this particular approach would say that, try as the modern Moroccan nation-state might to impose its borders and rules for crossing, it fails to bend all of its subjects to its will. This is particularly true of those who make a living crossing borders, namely, migrants and smugglers. Their tricks of the trade, their ruses and subterfuges, are developed hand-in-hand with the development of the borders meant to control them. Even though not on a par with the state's power, these acts of subversion do perturb the system. One could argue that they create counterhegemonic ripples that leave border peoples less enthralled with the grandeur and pomp of the state. This occurs in spite of the fact that many border people must run up against the state's borders and endure the harassment and humiliations of its agents on an almost daily basis.

Along the same lines, the resistance school might specifically draw attention to the way, for example, women smugglers succeed by using to their own advantage the patriarchal ethos undergirding the Moroccan state—its father-king ruler and its official religion—all represented at the border in the form of various male officials. The female smugglers could be said to be mocking these cops and soldiers and customs guards, defying them to drop their manly pretenses, to lower themselves morally to the point where they could physically search a female body. The women are betting that they will not.

The guards, by contrast, do extract *time* from the women as a price of passage by making them wait interminably. But time is one commodity that these poor women have plenty of, so, in effect, they can outwait the guards, who eventually grow irritated by their presence and pass them through.

Poor male smugglers also work to their advantage the traditional interaction style that commands obsequious behavior on the part of the dominated when in the presence of the dominant. Public tears, cries for mercy, supplicating hand gestures, bowing down or falling to one's knees in the face of authority—all eventually require a grudgingly merciful response. Open resistance and defiance trigger greater use of force, but supplication demands leniency. And so smugglers plead and cry and wail loudly and eventually crawl across the border, having resisted the state agents' attempts to prevent them from smuggling by, paradoxically, not seeming to resist at all.

I recognize how powerfully seductive this approach can be and how important the study of resistance is to anthropology (Brown 1996; Ortner 1995). Yet, I think that the approach, as it has been developed, by laying primary stress on the way everyday people in everyday life elude measures meant to control and contain them, misses the interesting process by which the dominated often become complicitous in legitimizing state power. I want to argue, against de Certeau, that subjects involved in border running can both succeed at what they are doing while simultaneously confirming the legitimacy of the state's monopoly on ceremony and violence—which, I think, is the true key to their success, not the various ruses they may deploy. In other words, it is not just that they have outwitted the state's agents, or resisted its laws, or pulled off a popular subterfuge that the border guards have yet to figure out. It is that they have met the symbolic requirements of subjects, namely, they have been forced, willy-nilly, to pay obeisance to state power. This is Mbembe's remarkable contribution to the study of domination and subordination. He, unlike perhaps any of the resistance school theorists, has shown how popular recognition of the state's legitimate monopoly on ceremony and violence is what underlies most of the interactions between agents and subjects of the state. That recognition is what the state seeks to wrest from its people.

Mbembe is not alone in recognizing that acts of resistance might be better understood as sites for the reproduction of state power. Enemies of the state who operate as if they were struggling against an evil Leviathan or a unified, powerful entity practice a kind of reification of the state that is no less effective than the work of public administrators, the police, or the army in propagating an image of state omnipotence. They all "fetishize" the state, to use Michael Taussig's (1992) term.

Taussig's elegant meditation on the state suggests that state power is reified most completely by those inhabiting the margins. They are most under the influence of the image of the state with a big *S*. They live most completely with "the fictions of will and mind thus inspired" by its sacred unity of reason and violence (1992:116).

But *fictions* of will and mind? Taussig leaves the impression that if citizens only understood how state power operates, if they could only see behind the mask of state ideology, they would be free of its hold over them. His metaphorical use of the image of the mask and his emphasis on the religious character of belief in the state suggest that a kind of false consciousness is at large in the land. Faith in the state is a blindingly powerful force that leaves people enthralled; nevertheless, it is a blindness that can be cured. Other commentators have suggested that a quasi-religious subservience to the state on the part of its subjects need not be present for state power to be reproduced in them. It is not a matter of demasking or of false ideology. In fact, state agents are usually well aware of the high level of skepticism and cynicism present in the people. It is what they do, not what they think, that matters.

WEDEEN ON COMPLIANCE AND CULTS OF PERSONALITY

Lisa Wedeen's (1999) study of the cult of personality that kept the Syrian state afloat under the late dictator Hafez al-Asad suggests that constant, orchestrated participation in rituals of obedience creates a compliance in state subjects that precludes the need for demonstrations of state violence. Such acts as displaying portraits of the leader in shop windows, portraying the leader's smallest deeds nightly on state television, throwing national birthday celebrations for his "Greatness," and reciting yearly mass oaths of allegiance all cause subject citizens to become complicit in their own domination. This is not a form of hegemony, as Wedeen (12) points out. Subjects living under cults of personality do not for a minute believe regime claims or think it a natural state of affairs. But they do end up having to display publicly their conformity with the regime's guidelines for acceptable speech and behavior. They are forced to exhibit an outwardly respectful posture. And the ability to impose that disciplinary regime upon its citizens is a kind of power. As Wedeen (12) puts it, the regime alone monopolizes the ability "to compel people to say the ridiculous and avow the absurd." So getting people to habitually repeat formulaic government slogans and to perform gestures of adoration—all within public spaces and discourses dominated by government-controlled rhetoric and symbols—constituted acts of obedience in Asad's Syria. Other mechanisms of coercion were not necessary. As Wedeen puts it provocatively, "The cult was itself such a mechanism" (145).

Wedeen's discussion of the ways the state makes its presence felt conjures up images that are much more banal than the illustrations of physical abuse I mentioned earlier. In fact, in reflecting on the case of Nador, officialdom often seems incapable of restraining itself or of measuring when its images and gestures have become stale. I am thinking of the scene across the bay from Nador where on a hill are located some whitewashed stones arranged so they spell out the three words of Morocco's official slogan, "God, Nation, King" (*Allah, al-Watan, al-Malik*). The slogan (similar rock logos ride astride numerous hills throughout Morocco) is visible from Nador because the Arabic letters are ten times larger than life. But, rather than suggesting a sinister, panopticon-like edifice, the crudity of execution and the symbolically excessive dimensions suggest, instead, that the giant slogan forms one of the sites where power "vulgarizes itself" (Mbembe 1992:129), that is, it spreads out and over the population in the process of becoming ordinary and unremarkable. Indeed, the challenge represented by the presence of this giant Arabic edifice in the midst of *Thamazight*-speaking country seems to raise few eyebrows among literate Nadoris—it has become so much a part of the landscape. The slogan, which could only be in Arabic, for that is the written language of the state, has the potential, in Wedeen's scheme of things, of reminding the locals that whoever controls writing controls the reins of power. In Wedeen's analysis, it could also be construed locally as a visual signifier of the dominance of the Moroccan Arab interior over the Berber-speaking margins. It is neither. It is not even received as a mundane civics lesson. Instead, my anecdotal investigations revealed that people thought of it, when they thought of it at all, as decoration. Like the presence in every city of a boulevard named "Hassan II," after the former king; and the spidery wrought iron gateways at the entrances to most smaller towns arching over the street to form a scaffold holding up either the official state logo, the symbol of the monarchical crown, or the nation's flag; or the obligatory back-lit photos of the king along the hundred meters or so of roadway leading up to every city limits in Morocco—like these, the stone logos also function to transform Moroccan routes into public arenas for the hackneyed display of majesty.

My sense is that political discourse and symbols in Syria are more under the control of the regime than in Morocco. The single party apparatus with its Leninist approach has no parallel in the Moroccan kingdom. Perhaps the Moroccan monarchy does not need such tight controls because of its greater legitimacy in the eyes of its subjects. Or perhaps it is just a matter of following different paths to regime stability. The bureaucratic authoritarianism of the single-party Syrian state enforces obedience and creates public powerlessness by requiring constant spectacular mass performances of that obedience (Wedeen:147). The Moroccan regime—at least in the border regions like Nador—seems to place more emphasis on ostentatious display coupled with petty, quotidian manifestations of violence.

MBEMBE ON CEREMONY AND VIOLENCE

Mbembe (1992) continues this line of argument, but makes it more provocative and more specific when he points out that state power in weaker, postcolonial nation-states like Morocco is characterized by the way it impresses itself upon its subject population by means of equal measures of buffoonish ostentation, carnivalesque pageantry, and eroticized pain and violence aimed at the body of its subjects (Mbembe 1992). In such a post-colony, the autocrat and his agents demand harsh adherence to the daily routines of obedience (posting of official photos, wearing of uniforms, etc.) and extract their due (in the form of taxes, duties, license fees, embezzlement, confiscation, bribes) with impunity from the subject population. They simultaneously dramatize the legitimacy of their authority to do this by means of extravagant rituals constructed from the symbols and images of power. The official line always stresses the unity of the subjects' adoration of their divine leader. The image of an obedient populace receiving the benevolent gifts of the ruler is repeated in various official forms, over and over again. The very excess so often entailed in state ceremony itself feeds back to reinforce the legitimacy of the system, for only those who dominate are capable of such extravagance. Thus, as Mbembe writes, "officialdom and the people share many references in common, not the least of which is a certain conception of the aesthetics and stylistics of power, the way it operates and the modalities of its expansion" (1992:13). Wedeen, I believe, would recognize this description and agree with its premises.

But the other half of Mbembe's equation states that the postcolony, as personified by the autocrat and his henchmen, binds subjects not only through ostentatious, symbolic displays, but also through the mechanisms of institutional violence. The swift and sure application of painful procedures elicits from subalterns personal and intimate awe and honor as well as dread. In short, the state habituates subjects to subjection through daily practices of injury and injustice.

Let me back up and try to tie Mbembe's ideas to the scene at the Moroccan border. He suggests that just such petty forms of the exercise of authority "tire out" the bodies of the subject population made to tolerate the burden. The effect, as he puts it, is "to 'disempower' them and to render them docile" (Mbembe 1992:13). Smugglers and other subjects come to regard the cuffs and kicks, the snarling, and the indifference of the border guards as being among the prerogatives of power. They mutually share with these state agents the conception of these acts as the ways power manifests itself. The guards indulging in what to the Euro-American anthropologist appears to be excessive, if often petty, force are in fact best understood as performing domination. They are publicly displaying, on the bodies of the smugglers, the ruler's right to rule.

Let me put my point another way: I mentioned a moment ago that I thought subjects involved in border running can both succeed at what they are doing while simultaneously confirming the legitimacy of the state's monopoly on ceremony and violence. More precisely, I think smugglers succeed at what they are doing *because* they are complicitous in their acceptance of the regime's right to rule. In the Nador-Melilla border case, I think this comes about via various obsequious expressions of fealty extracted from smugglers by Moroccan state agents as their price of passage across the Spanish-Moroccan border. The agents are not there to prevent border passage—smuggling or otherwise—but to require a show of recognition of the state's right to dominate. They are not "taken in," as de Certeau might have it, by smugglers feigning helplessness; they are not being bluffed by smugglers pretending to throw themselves on the mercy of the guards. On the contrary, such public acknowledgement of state power is exactly what the guards, as agents of the state, are seeking. The border is thus more of a stage upon which the state's violence and pageantry play themselves out. The numerous checkpoints and barricades, the multitude of uniforms, the rows of flags, even little details such as passport stamps and so forth—all of these testify to the power of the state to command its subjects. What is more, the symbols of this apparatus of domination are meant to instill in subjects a habit of obedience to the state. Thus the wailing for mercy, the pleading on bended knee for leniency, and the beseeching of guards for clemency are better explained not as de Certeau–style subterfuges, but as public signs of the recognition on the part of smugglers of their subservience, and as demonstrations of their acceptance of officialdom and its natural right to authority. That public demonstration, in the end, I would like to argue, is the price of passage extracted by border guards from the smugglers.

NOTES

1. An earlier version of this chapter appeared in *In and Out of Morocco: Smuggling and Migration in a Frontier Boomtown* by David A. McMurray (Minneapolis: University of Minnesota Press, 2001); reprinted with permission. The map of Morocco was created by Bill Lanham. All photographs were taken by the author.
2. Apparently, this is not all that uncommon a position for people living in borderlands. See, for instance, Flynn (1997:324).
3. Rifis acquired colonial names for the first time when they went to register with the authorities to get identity cards in order to cross to Melilla. Unlike the French, the Spanish had not pursued a vigorous policy of registration, so many Rifis had yet to acquire a "colonial" name at the end of the Spanish Protectorate.
4. Ironically enough, though, the major story recounted about the assault is that the king was winged on the nose by a Rifi bullet or knife during the battle. Look at any of his portraits in any official building, say the locals proudly,

and you will see a scar across the bridge of his nose. That is a souvenir of King Hassan's first and last visit to the Rif.

5. Remember, these are figures for the "usual suspects" rounded up. I am guessing the statistics comprise mostly novices or out of towners who do not know how to play by the rules. The vast majority of smugglers are merely shaken down and then sent on their way, or pass through back exits, or cross by sea.

6. Student anger at the shrinking number of suitable white-collar job opportunities continues to grow. The same goes for their resentment of credential inflation. A bank employee in Nador told me that, when she started in 1980, some lycée education was sufficient to obtain her job. By 1986, the same position required a university degree, related course work, job experience, and knowing someone in authority.

7. Popular memory of the riot in Nador seemed to highlight the helplessness of the inhabitants in the face of the arbitrary brutality of the repression. Locals claim that the Arab troops responded so savagely because they had been told that the northern Berbers were trying to secede and declare an independent state and were therefore as much traitors and enemies of the nation as the Polisario, the enemy most recently faced by many of the soldiers. Ethnic tension thus intensified as a consequence of the riot. In addition, students and teachers now live in an atmosphere of paranoia and enforced apathy, and no student demonstrations have taken place since. Prior to that, student demonstrations were said to be practically monthly affairs.

8. We met a young man from England who had bought a couple of kilos of hash in Melilla and tried to smuggle them back to Europe but was caught. He said it cost him two months and 5,000 British pounds before he could get the Spanish judge to release him.

9. A U.S. importer of Moroccan mirrors told me that he had been approached by hash smugglers in Casablanca who wanted him to place the drug in the back of the mirrors. He claims he refused, but said it would be easy enough to do because the New York customs agents inspect containers on the first three or four shipments, and then, after that, leave shippers alone.

BIBLIOGRAPHY

Amraoui, Mohamed
1993   Rif, La guerre au kif. Maghreb Magazine no. 15 (June): 28–40.
Barjila, Abd Al-Salam
1985   Al-Nishat Al-Tijari bi Madinat Al-Nador. Fes, Morocco: Université Sidi Moh. Ben Abdallah, unpublished mémoire en géographie.
Brown, Michael F.
1996   On Resisting Resistance. American Anthropologist 98(4): 729–749.

Certeau, Michel de
1988   The Practice of Everyday Life. Steven Rendall, trans. Berkeley: University of California Press.

Clément, Jean-Francois
1986   Les Révoltes Urbaines de Janvier 1984 au Maroc. Réseau villes Monde Arabe (Mouvements Sociaux) no. 5 (November): 3–44.

Flynn, Donna K.
1997   "We Are the Border": Identity, Exchange, and the State along the Bénin-Nigeria Border. American Ethnologist 24(2): 311–330.

Kearney, Michael
1991   Borders and Boundaries of State and Self at the End of Empire. Journal of Historical Sociology 4(1): 52–74.

Mbembe, Achille
1992   The Banality of Power and the Aesthetics of Vulgarity in the Postcolony. Janet Roitman, trans. Public Culture 4(2): 1–30.

Noakes, Greg
1993   Morocco Declares "War on Drugs." The Washington Report on Middle East Affairs 12(1): 56, 85.

Ortner, Sherry B.
1995   Resistance and the Problem of Ethnographic Refusal. Comparative Studies in Society and History 26(1): 126–166.

Ouariachi, Kaïs Marzouk
1980   Le Rif Oriental: Transformations Sociales et Réalité Urbaine. Thèse de troisieme cycle, Ecole Des Hautes Etudes en Sciences Sociales, Paris.

Poulantzas, Nicos
1980   State Power, Socialism. London: Verso.

Slouschz, Nahum
1927   Travels in North Africa. Philadelphia: The Jewish Publication Society of America.

Soudain, François
        Guerre à la Drogue! Jeune Afrique 1670 (January 7–13): 16–20.

Taussig, Michael
1992   The Nervous System. New York: Routledge, Chapman and Hall.

Wedeen, Lisa
1999   Ambiguities of Domination: Politics, Rhetoric, and Symbols in Contemporary Syria. Chicago: University of Chicago Press.

CHAPTER 6

REPRESENTATIONS OF CRIME
On Showing Paintings
by a Serial Killer

Anne Brydon and Pauline Greenhill

IT IS OBVIOUS THAT A CRIMINAL'S CRIMES affect his or her victims. It is less clear how an offender's noncriminal actions can affect the victims of another's crimes. But when Winnipeg, Manitoba, Canada's Plug In Gallery, an artist-run contemporary arts space, proposed to display three paintings by executed American serial killer John Wayne Gacy,[1] the potential for extension of a criminal's abusive power by the public presentation of apparently nonviolent visual imagery became a matter of heated debate.

The media and public dissection complicated this situation of crime and its implications with another discourse around the injustices perpetrated upon both art and society by the censorship of artistic expression and the media exploitations of social controversy and violence. Other commentaries pitted victims of violence against aesthetes, and the common folk against the elite. And many arguments circled around the symbolic, evocative, and even magical connections between the creators of art and their products.

As Henry Giroux (1994:202) has pointed out, "the struggle over identity can no longer be seriously considered outside the politics of representation." At issue here are several identities—including citizen, victim, artist, and curator—and the ways in which conflicts reify and balkanize them. The actions of key figures in the debate became gambits in a public contest for moral superiority. Cultural capital accumulates on the side of the victor.

From the time the planned exhibit was publicly announced until the show closed, regardless of the fact that the show included works by over 20 artists from five countries, media attention, as well as most public and much private discussion, remained firmly fixed on one question: To show or not to show Gacy's work?

But despite this media construction of diverging constituencies (artists and curators versus victims and folks) and perspectives (freedom of expression versus censorship; literal versus symbolic interpretations), neither the quotidian locations from which identity politics were argued nor the actual discourses were simple or dichotomized. Much public and private commentary implicated contingency and negotiation, engendering what Homi Bhabha (1994) calls a "third space" in which new meanings and resistant messages can be mobilized from dominant signs and symbols. Indeed, it might have been possible to turn what eventually became a theater of the absurd against the tabloid press by nurturing the strong potential for collaboration and coalition between two embattled constituencies, both threatened by social and fiscal conservatism—artists and victims of violence. But such an alternative failed to happen. The third space remains nascent, awaiting, perhaps, more favorable circumstances.

Our ethnographic research sheds light upon these altercations by placing them in several wider social and cultural contexts and by including a place for conversations, formal interviews, and other nonmediated discourses. These sources, and our experiences as participants in and observers of Winnipeg's cultural life, help to show the constructedness, but also the multiplicity, of viewpoints, and we refuse to polemically centralize or approve one or another perspective. We seek to bring together productively the evident cacophony of voices speaking to and about the issue of showing the Gacy works in Winnipeg, many of which have hitherto been interpreted as divergent. We also explore the possibilities of an art practice that cuts across discursive boundaries defining a range of constituencies—artists, curators, victims, viewers, and the general public.

Making an ethnography that traces the effects of a highly fraught image complex such as "the serial killer" requires the tools of discourse analysis in order to capture the dynamism of its translation into different social networks. A more conventional anthropological approach on a more conventional topic would attempt to explain the Gacy controversy with reference to a single, stable social context. However, in this case, where a mass-mediated image complex enters into a local population made up of a myriad of discursive practices also mediating between the so-called local and global, attempting to make a distinction between foreground and context loses the subtle ways in which one constitutes the other simultaneously. Discourse analysis enables examination of meaning-making as social and cultural practice. We, the authors, use it as part of our ethnographic study of how a criminal figure, "the serial killer" (which is at once real and socially constructed), mediates the production of social identities in a politicized debate over power and representation.

We interpret the Gacy controversy from a general concern with the double meaning of the term "representation." In one sense, it invokes notions of public good, and who can claim the right to represent common interests. Is the identity of victim sufficient to speak for all victims? Is the identity of artist sufficient to speak for all artists? Should curators decide what is art and who are artists? Do all artists, and all victims, have a common identity/body politic?

In another sense, representation raises questions about how people interpret visual messages in general, and painting in particular. How does curating shape these perceptions? Is a work of art's meaning determined by the sum of its materials? Should consideration be paid to the experience and knowledge viewers bring with them? How are art and artist personally and politically linked?

THE PROBLEM

The three Gacy paintings belong to Glen Meadmore, a Winnipeg-born musician and performance artist now living in Los Angeles who had commissioned four portraits in 1988 and 1989 to use as cover art for a planned CD of gay Christian rockabilly.[2] They are based on photographs given to Gacy, depicting Meadmore in various personae.

Only one painting is titled. "Glen Riding High" is the only image that could be called a "banal circus painting," a phrase many we talked to typically used to describe Gacy's pictorial themes, and one which does not capture either of the two other paintings. The placement of a clown face covering the genitals of the male figure in another painting is hardly innocuous, even without the knowledge that Gacy once performed as Pogo the Clown at hospitals and children's parties. The third work was reproduced in *MIX* magazine with the caption "one of Gacy's circus paintings" (Mitchell 1997), although it is clearly not that. A figure in a death-head T-shirt leans on something we initially interpreted as a torso with severed legs. Meadmore says it is a giant penis.[3]

On October 26, 1996, Plug In Gallery launched its newly renovated space by staging an extensive group show called *The Moral Imagination*. Gallery curator Wayne Baerwaldt, never one to shy away from provocation or controversy, had planned to include the three Gacy paintings, along with works by over 20 locally, nationally, and internationally reputed artists.[4] The eclectic mix of artworks, in Baerwaldt's view, addressed "the moral imagination," which he described in the following terms:

> The moral imagination can be a questioning of how images and other artworks communicate a moral sensibility (or do they?), how their creators are invariably bound to a commercial image world that mediates and often obfuscates important insights into how the artist intended the work to operate. . . . [It] is also about "detailing the processes by which the subjectivity of objects is produced in human experience" (W. J. T. Mitchell, "What Do Pictures *Really* Want?"—October #77, 72), so as to suggest that pathological symptoms of the artist or a subjectivizing of images becomes relevant and occasionally dominant in the viewer's mind (Baerwaldt 1996:1–2).

Baerwaldt's essay, produced in *The Plug In Harold* that accompanied the show's opening, raised these issues of commercial influence, artistic pathology, and intention, particularly focusing upon media sensationalism of horror and evil.

When news of the planned exhibit became public in mid-September, Winnipeg's tabloid media vilified not only Baerwaldt and Plug In, but also the governmental arts funding programs on which his curatorial work and gallery depend. CBC radio's *As It Happens* interviewed Baerwaldt, giving the controversy national coverage. Open-line radio shows and continued press coverage kept the issue alive and local public opinion engaged until *The Moral Imagination*'s opening five weeks later.

Plug In made at least one attempt to address some of the concerns raised. They hastily convened a meeting on September 23 so that Plug In members, artists, and representatives of Citizens Against Violence could discuss the issues. That latter group's president, Beverley Frey, said "I do not think that these works should come to Winnipeg. There are so many victims of pedophilia; it's unfair to children and parents to see Gacy's paintings *in the papers*" (our emphasis). Member Tracey Walsh stated that "the paintings by Gacy should never have been brought to Winnipeg. They were painted by the same hands that murdered and tortured. I beg you not to show the work. . . . Take this crap out and bring in another deserving local artist" (both quoted in Plug In minutes 9/23/96).

In response to the public outcry and to administrative and funding problems that could be raised by showing the work, Baerwaldt changed his plans. He told one reporter, "I just don't have the right conditions to show

6.1. As the *Winnipeg Sun* stirred objections to displaying the art of a serial killer, it became the only forum through which the public could view his images.

the work in the proper light," but would not specify the conditions he lacked (St. Germain 1996c:5). Later, he called his withdrawal of the paintings "a political, internal matter. It became impossible for me to show them without certain things happening internally that I can't discuss" (Plug In forum, 11/27/96).

Instead of showing the actual works, Baerwaldt framed and displayed two pages from the tabloid Winnipeg *Sun*. On September 17, that paper's front page had a full-color reproduction of "Glen Riding High" with the headline "Gacy's Art Stirs Anger: Killer's Paintings on Display [*sic*] in Peg" (see figure 6.1). (In fact, the paintings were not at that time—or ever—displayed at Plug In; nor, arguably, were they ever displayed in the city, *except* on the pages of the *Sun*.) On page three, a second painting (also illustrated in *MIX*) was reproduced in color accompanying the headline "Local Gallery to Exhibit Killer's Death-row Art: Gacy Paintings Anger Victims of Violence" (see St. Germain 1996a). As Winnipeg *Free Press* art reviewer Garth Buchholz later noted: "We protest the paintings being hung in the gallery, yet we accept that they are reproduced in color on the front page of a newspaper where even more people see them" (Buchholz 1996:B4). (Actually, Citizens Against Violence's Bev Frey did denounce the paintings' appearance in the papers at Plug In's meeting, as noted above.) The controversy proved to be good advertising, and *The Moral Imagination* ran an extra two weeks to accommodate the higher than typical attendance.

In our view, Baerwaldt's *Harold* essay, and his other public comments, failed to illuminate the issues, engaging instead in an obfuscating and difficult postmodern rhetoric around representation, or singling out the media as (re)creators of moral ambivalence. The debate quickly polarized to pit freedom of expression against censorship or, alternatively, upstanding community values against degenerate parasitism on the public purse. Although people both inside and outside Winnipeg's arts community expressed more nuanced opinions that did not align them with either position, these views were lost in the clamor to gain the moral high ground.

FREEDOM OF EXPRESSION VS. CENSORSHIP: INVOKING THE AVANT GARDE

Some supporters spoke sympathetically about the need to consider uncomfortable topics in a public display. Others, even those unable to grasp the purpose of displaying Gacy's paintings, were willing to consider it reasonable. Most artists with whom Brydon spoke—but not all—interpreted the controversy in light of Baerwaldt's typically provocative curatorial style.

Many considered his actions careless, even reckless, and lacking substantive intellectual purpose, but having potential negative repercussions for the entire arts community. Some took delight in seeing Baerwaldt caught out in a publicity stunt of which he had lost control.

Baerwaldt's critics, however, soon felt compelled to remain publicly silent. To speak out could be construed as disloyal to the arts community as a whole. Nonetheless, Cliff Eyland, artist, curator, art writer, and vice-chair of Plug In's executive board, wrote disparagingly in *MIX* magazine about artists who "had objections to the show that scarcely differed from those of the tabloid press" (1997:22). He defended Baerwaldt's curating as instinctive, arguing that the lack of a clearly voiced curatorial premise did not undermine *The Moral Imagination*'s potential effectiveness to fulfill Plug In's mandate of addressing controversial topics and pushing public opinion and awareness.

Despite these mixed sentiments among artists, public rhetoric from art-scene commentators relied on the trope of freedom of expression versus censorship to narrate the Gacy controversy. Richard Noble, at that time a professor of political theory at the University of Winnipeg, reviewed the show for the arts magazine *Border Crossings*. He dismissed the pre-show controversy as "drearily predictable" and playing out a "well-rehearsed process" (1997). He denied the authenticity and credibility of Baerwaldt's critics, precluding consideration of their arguments' moral weight. He attributed the outcry to a general inability to understand art: "Lacking a public language in which to discuss the moral implications of controversial art works, censorship is invoked immediately" (19). Noble also interpreted the paintings' withdrawal not as a moral choice, but as an act of self-censorship—a capitulation to public opinion. He failed to interpret the paintings in question in any way that might have enlightened the audience as to why they should have stayed.

Noble's judgment begs several questions about the relationship between freedom of expression and moral choice, and between censorship and choice of any kind. For instance, one might ask whether freedom of expression is an absolute value overriding all other considerations; withdrawing the paintings could be an act of social responsibility. Noble did not address these issues. Instead, choosing to focus on individual works by other artists in *The Moral Imagination,* he implied that the presence of art that he read as making moral statements vindicated the entire show: "It is to Plug In Gallery's credit that they have organized an exhibition which brings into such sharp relief the need to develop a better public language for talking about images" (19).

Noble also organized a public forum at Plug In during the show's November run, with Noble himself, Wayne Baerwaldt, Morley Walker (entertainment editor of the Winnipeg *Free Press*), and Sharon Alward (performance artist and professor of art at the University of Manitoba) as panelists. Noble's flyer billed the event as an opportunity "to discuss a number of issues, including censorship, arising out of . . . *The Moral Imagination*." But in selecting participants, choosing the locale, and framing its subject matter, he effectively directed the discussion toward criticism of media coverage of all arts-related activities. Rather than considering Baerwaldt and curatorial practice, the forum actually protected him by directing attention and blame toward the media and away from divisions within the arts community. With only a few exceptions, the arts-dominated audience placed Morley Walker in the hot seat, making him defend his newspaper's arts coverage. Ironically, the *Free Press* had attempted to balance its coverage of the Gacy controversy, unlike the Winnipeg *Sun,* which had sensationalized it.

In order to discuss questions of artistic freedom, writers such as Noble and Eyland implicitly situated the Gacy controversy within the historical logic of avant-garde art. This rhetoric framed the understandings even of critics speaking privately. They thus missed an opportunity to engage in more contemporary debates around art's potential and actual social meanings. Instead, Gacy was linked with artists whose imagery has provoked public outrage, such as Eli Langer (a Canadian who successfully challenged in court the police confiscation of his pencil drawings depicting children engaged in sexual acts), Robert Mapplethorpe, and Andres Serrano (cf. Becker 1994).

At the heart of this association between Gacy and the avant garde rests the acceptance that shocking the bourgeoisie is a form of social commentary, champions freedom of expression, provokes social change, and challenges middle-class hypocrisy. Without identifying the problem *The Moral Imagination* was to address, or evaluating the effectiveness of the planned Gacy inclusion, many invoked art history as if it provided an immutable moral truth legitimating Baerwaldt's curatorial action.

Eyland's and Noble's (and others') comments treated public reactions as a mysterious misapprehension of art, rather than as a visceral indication of engagement with the meaning and power of representation. Their responses are too compliant. They fail to consider that Gacy was neither avant-garde nor (indisputably) an artist. They assume that any art possesses absolute moral value that transcends audience perceptions. More crucially, they fail to address their own underlying moral relativism, that productions undertaken in the name of art may be judged differently than similar ones in other contexts.

That showing Gacy's paintings—or, indeed, any curatorial action—can be avant garde is debatable. In part initiated by the French Realists, the avant garde arose in the mid-nineteenth century as part of artistic modernism. Modernism assumes that good art is new, progressive, and ahead of its time, and that artists can act outside history and must elude social constraint to explore truth in art-making. Ideally, a true avant-garde gesture combines tactical shrewdness with elegance and wit, flashing up in a moment of flux and disappearing when the moment passes. But over the years, its shock techniques have become institutionalized in the art world and have thus lost much of their original import. The valorization of the artist as outsider and privileged observer—a modernist mythical figure paralleling that of the anthropologist—misses how bourgeois values appropriate avant-garde gestures. The late-twentieth-century middle-classes are not rigid and condemning protestants employing a nineteenth-century work ethic. Shock now sells products,[5] draws movie audiences, and provides fodder for daytime TV chat shows.

For artists, claiming to belong to an avant garde has become a means of creating social identity as much as describing artistic practice. This is not surprising: Those making thought-provoking art are, with few exceptions, marginalized, as their meager incomes and extensive labors to find an audience testify. Some need a socially recognizable identity to provide a safe psychic space from which to make art. Yet, an artistic bohemia, a creative social space, is distinct from the avant garde, a social identity that privileges its views over others. As in the creation of any group, boundaries defining insiders of the avant garde also act to exclude and diminish outsiders who fail to understand its codes and mores. During the Gacy controversy, critics who did not privilege or acknowledge avant-garde norms were dismissed by art commentators.

The will to defend Baerwaldt's curatorial practice through a display of arts community solidarity against censorship, as imperative as it seemed to some, overlooked alternative strategies. Art historian Leslie Korrick, an audience member at Plug In's forum, commented that the media could actually have been a useful means for creating public discussion:

> A situation was created whereby the choices were either the paintings stayed in the show, or the paintings were eliminated from the show. I am wondering if there isn't a creative third option that might have been pursued so that the paintings, which you initially felt were significant and needed to be in your show, could have remained and, at the same time, there was room for various audiences in the city to respond to them. . . . Maybe there are more radical ways of presenting the work that don't rely on the media to carry your message forward. I wonder if just sitting back and hoping for the best, in terms of what the media produces . . . allegedly on your behalf, is all that helpful.

In an interview with Pauline, Winnipeg artist Grant Guy, whose work was included in *The Moral Imagination,* suggested one such alternative. He

commented that Baerwaldt should have done "more homework" in order
to involve a variety of constituencies:

> Deal with questions of art and insanity and morality, art and prisoners' ther-
> apy. . . . There should have been criminologists involved, victims' groups,
> police, prisoners' rights people . . . presentations and things building up to
> it so that there's a context, not all these red herrings that Plug In was forced
> to screen out because they found their backs to the wall. . . . If Plug In would
> have worked the show through in a more collaborative effort with the com-
> munity, the *Sun* would have been castrated from the beginning (1997).

Dealing with issues around prison art and outsider art (e.g., Maizels 1995,
Beardsley 1995, Zolberg and Cherbo 1997) could acknowledge Gacy's significance without relying exclusively on sensationalism.

Alternatively, artist Sharron Zenith Corne emphasized promoting social
understanding over the risks of censorship:

> If you think about the holocaust and how many people were killed and
> damaged by it, [what would happen] if it had been put under the rug as
> opposed to seeing photographs of it? . . . I think *Mein Kampf* isn't a bad
> parallel. It is a book, and maybe that makes it a little different than a pic-
> ture. But nevertheless it is in all libraries. . . . I wouldn't think of that as
> glorifying it. . . . One of my historian friends . . . said "I think it is impor-
> tant that people do read it and learn from it." . . . It does set up a kind of
> mystique when suddenly you know these things are there but they are not
> accessible. . . . Suddenly they gain another kind of status which makes
> them important in a different kind of way (1997).

Thus, if Plug In had led a discussion that acknowledged the power of
the Gacy works, this could potentially have created a space for reflection
rather than reaction. Victims of violence might then have understood the
display of the works of a perpetrator of violence as extreme expressions of social and psychological dislocation and disorder, rather than as a glorification
of the offender.

VICTIMS VS. ELITES

The tabloid media represented the conflict around Gacy's work in highly dichotomized and reductive terms of class politics. On the one side, they said,
were the elite—curators, academics, and artists whose actions were based in in-
tellectualized, ivory tower notions of art and morality. Winnipeg *Sun* colum-

nist Naomi Lakritz called them "The sandal-shod habitués of the local gallery scene, steeped in their highbrow pretentiousness" (1996). The obverse in this tabloid creation was the public: victims' rights groups and other real people who experienced their lives directly rather than filtering them through ideas.

A Plug In employee commented:

> The media weren't interested in us saying, "You know what? We think we made a mistake here. We're going to take these paintings down," . . . even though that is what they presented as wanting. . . . For them . . . it has to go "Rant, rant, rant." Just keep going back and forth as long as possible, and see how much blood would get spilled (Duggan 1997).

In "Tax bucks for sick art irks public" (St. Germain 1996b:3), *Sun* reporter Pat St. Germain linked showing the Gacy paintings to Medicare reductions by interviewing the otherwise unidentified "hospital worker Lee Rempel": "'If it's privately funded, that's fine,' [Rempel] said of the non-profit gallery's plan. . . . 'But all three levels of government are paying for this. Taxpayers are paying to have this mass murderer's artwork displayed'" (ibid.). In this reading, money that should have gone to saving lives was instead being squandered on irresponsible and arrogant artistic poseurs. These are the kinds of stereotypes and prejudices about the arts against which the Canadian art world must continually fight in order to prevent further erosion of credibility and funding. Not surprisingly, the Winnipeg arts community responded defensively to this free enterprise rhetoric.

Bolstering this compelling conspiracy of government-sponsored money-wasting, St. Germain quoted Bill Lobchuk, owner of a private gallery, who "said the Gacy exhibit is just the sort of outrageous stunt that's helped to make Plug In 'the darling of the funding agency boys . . . It's got nothing to do with art, it's just publicity'" (ibid.).

For their part, Winnipeg's Citizens Against Violence were prompted to speak when St. Germain called them for comment—in itself an act of constituency-making. The group publicly speaks for those silenced or ignored when violence has entered their lives. President Beverley Frey commented to me: "No one will listen to the victim half the time. . . . They just . . . say, 'Oh don't worry about it, they'll get over it.' But they don't, [not] for a long time" (Frey 1997). Many from the group, like Frey herself, are family members, while others, like Tracey Walsh, are victims themselves.

St. Germain evidently expected Citizens Against Violence not only to represent, but also to speak for, the perspectives of the families and friends of Gacy's victims. Yet her public responses spoke of personal experiences and effects, and, in conversation, Frey spoke emphatically for herself, but carefully considered alternative points of view:

> To me all the art should have been destroyed. No, I am wrong, because people will say "When I look at this art I am trying to figure out what is in this man's mind." . . . Maybe [Gacy's work should be] in a psychiatrist's office . . . [who] is trying to rehabilitate . . . the same type of person. . . . Or maybe they should have an art gallery for prisoners' paintings. . . . And if somebody wants to see a madman's paintings, they can go to that particular gallery. . . . And . . . maybe we should be writing one of the victim's parents and asking them how they feel (1997).

Although she doesn't frame this commentary in terms that curators and other arts workers would see as part of their common discourse, she is clearly concerned for the appropriate contextualization of the works. And, in an interview with Greenhill, Frey's "third space" involved artists:

> We have very good artists in this city. We don't need to bring somebody's pictures in who has done that. . . . This is really wrong, to make this as a memorial for this person who has done so much damage to so many. . . . I mean, how dare they? . . . We have a lot of good things in this city without bringing in some trash like that (ibid.).

Evidently, the concerns of Citizens Against Violence were primarily with the possible legitimation of Gacy should his paintings be shown alongside those of recognized artists. This legitimation and its implied indifference to victims' sufferings, perhaps more than the paintings themselves, were the sources of their concerns about harm to victims of violent crime and their families.

By dismissing those who spoke against the inclusion of Gacy's work as "vocal minorities" (quoted in Bray 1996:C1), Baerwaldt participated in the same divisive rhetoric the *Sun* used to marginalize public arts. His comment also ignored the ambivalence among artists and curators, which Main/Access Gallery's Stephen Phelps's opinion piece in the Winnipeg *Free Press* demonstrates:

> The one thing that ultimately condemns their display is this: the inconsolable grief of those who lost sons and brothers to Gacy's murderous rampage. Offering this fiend a posthumous showcase, however insulated with curatorial authority, surely heaps unconscionable insult to injury (Phelps 1996:A11).

Local artist Michael Olito, the only person to withdraw work from the show, disagreed. Although he is adamantly opposed to censorship,

It became obvious to me that whether I approved of Gacy or not, the show was about Gacy and nobody else. . . . As I put it to Wayne . . . he already had his circus and didn't need another clown. . . . When people started objecting to exhibiting Gacy's art, I think the way to go would be to show just the Gacys and really make your point in spades. . . . I have no problem with the Gacy paintings being exhibited especially in the context of a show that was confronting the publics of this work and saying, "Respond!" . . . I have a big problem with them being removed and I have a problem with asking other artists to contribute to the show (1997).

Winnipeg's Grant Guy said that he would have withdrawn his participation in the show if the Gacys had remained because he did not want to be "a bit player in a badly written farce" but felt that the decision about showing the works was:

Wayne and the Board's and the membership's decision. . . . The only decision I had to make was whether or not I'd be part of the show. . . . I asked Plug In that, . . . in the deliberation and execution of the decision, they use some compassion and courtesy in their final result. Even if they went ahead with it, I thought that compassion and courtesy were still required . . . I think sometimes [artists] should kick things in the ass. I think art should go after the people that really deserve it. I think we can be as ugly and mean . . . as we have to be to get to that target. But to start hurting people unnecessarily who have already been hurt . . . is too excessive (1997).

For artists, no necessary connection existed between concerns about the Gacy art and wholesale dismissal of Baerwaldt and Plug In; nor was remaining in *The Moral Imagination* connected to supporting serial killers. Artist Sharron Zenith Corne commented:

My first reaction was that I may have to pull out [my] work. . . . These people had undergone all this trauma, and they had a right to make that kind of decision, and we should be supporting them. . . . And then after rethinking it, . . . I decided that probably it made more sense to bring it out into the open. . . . And also a friend of mine, having been a victim of rape, told me that she thought it was very important for women to understand the sociopathic mentality. . . . She herself had been totally charmed by somebody, and had no idea that this person was going to violate her. So it made sense to open up the topic and really look at it, rather than just shove it in the closet and hope that it will go away (1997).

Winnipeg artist Szu Burgess, who strongly supported Baerwaldt's position, approached showing the work from a feminist and anti-homophobic perspective, and arrived at a similar conclusion:

> I was glad my work was right beside [the Gacy display]. I was really happy
> about that. It's interesting, too, because I ended up having discussions with
> people who thought that as a feminist I shouldn't be excited by having my
> show with a mass murderer's as a good thing. . . . And, too, as a homosex-
> ual person, the controversy around the fact that he sexually assaulted and
> killed boys. . . . So I'm not supposed to participate in that either. I felt that
> those kinds of reactions were totally knee-jerk (Burgess 1997).

Although many artists and members of the public spoke out against censorship, many also refused to limit the definition of the controversy. Board
member Sig Laser conceptualized the debate in communitarian terms of the
fundamental need for mutual consideration between people:

> The [women from Citizens Against Violence] were expressing real emo-
> tion and upset. . . . And in the face of that, I viewed it as an act of gen-
> erosity [to remove the Gacy paintings from the show, not] as
> self-censoring. . . . I view that as being part of this community in a lead-
> ing and provocative way (1997).

Such deliberations raise questions about the responsibility of the arts to
society. The Gacy controversy gave the *Sun* another opportunity to trash the
arts community, but Plug In's own focus on censorship versus curatorial freedom failed to change the discursive terms, to examine the underlying causes
of hurt, or to consider why art might be implicated in it rather than offering a commentary on it.

A curatorial strategizing of the audience's readings could have illumi-
nated the questions of morality in art that *The Moral Imagination* was intended to address. Instead, Baerwaldt's curating lacked what Pierre Bourdieu
(Bourdieu and Haacke 1995:21) calls "true symbolic efficacy" such as could
have been created by locating a community of interest between two marginalized groups—namely, artists and victims of violence.[6]

PAINTED BY THE SAME HANDS
THAT MURDERED AND TORTURED

Censorship and freedom of speech were not the only narrative strategies
found in art discourse. Baerwaldt and his defenders deflected criticism onto
the chimera of victims' rights groups as extreme, powerful, and inflexible. To
do so, they mobilized the Enlightenment or modernist trope that divided
constituencies into rational and irrational. Eyland claimed, "We are being
held hostage by victims' rights groups" (Plug In minutes 9/19/96) and said

that the controversy had plunged the arts community into a "moral panic" (Eyland 1997). Baerwaldt complained that his attempts to discuss the show with Tracey Walsh and Beverley Frey were stymied by their inflexibility and lack of interest in his curatorial premise:

> "My name is Deborah, I am a victim." That's it and then you go on from there, and it's your stage, and there is no interest in the art work at all (Plug In forum 11/27/96).

Rather than interpreting Walsh and Frey's concerns as indicators of differing moral values, or engaging the women in further discussion, some arts commentators speculated about the victims' rights group's motivations. They represented Citizens Against Violence as incapable of perceiving an inchoate moral commentary invoked by showing Gacy's paintings. Despite Beverley Frey's comment, cited earlier, that "it's unfair . . . to see Gacy's paintings in the papers," they pilloried the group and the individuals representing it as hypocritical because they were seeking media attention while criticizing the gallery's plans as sensationalist and media-seeking.

Yet the symbolic connections between art and artist, and between art and audience, that create value and meaning are central not only to the discourse of Citizens Against Violence, but also to academic understandings of art. When *As It Happens* asked him about the motivations of the paintings' collector, Baerwaldt characterized him and the general public as weirdly fascinated by violence:

> Certainly, it's a form of titillation, that sort of fascination that, of course, started with Edgar Allen Poe in the 19th century where he detailed . . . this sort of stalking of a murderer, the aftermath of a murder. So there's been an audience for this. And I think all of us in our enthusiasm for bloodthirsty North American hockey and Hollywood films—as blank as they are—there is that strange invisible terror that we accept more and more. And I question whether we should; in some cases. . . . We [at Plug In] are interested in . . . this notion that audience members bring a weird sort of creepy alter ego to viewing these pieces. They're curious, they're fascinated. And artists have long been fascinated with that ability for freedom of expression to the nth degree.

In conversation, Baerwaldt clarified this last, apparent non sequitor. Expressing fascination with the idea of being outside society, he argued that great artists break from precedent to create new forms from a position outside society. Gacy, he said, was similar in that his murderous acts abandoned all social mores and emerged directly and unmediated from his desires. The other work in *The Moral Imagination* required the viewer to interpret the artist's codes, metaphors, and allusions via the art objects themselves, which were the primary linkage between viewer and artist. The Gacys required a different cognitive frame—knowing about the maker's identity, rather than knowing about art making. The victims' rights people, then, seem justified in supposing that Baerwaldt considers Gacy's murders to be his true art and his paintings the epiphenomenal by-products.

Cliff Eyland's *MIX* essay juxtaposes two constituencies: those who perceive the gallery as a temple, and those who consider it a courtroom. Should art be worshipped as sacred and kept inaccessible as the right wing wishes? Or should it be viewed dispassionately and judged "as if the art were on trial"? Eyland appears to advocate the latter, presupposing art viewing as a cerebral process, without emotion or sensation, and the courtroom as a neutral place of objective laws and empirical rules of evidence.

While critics state that art discourse's obscurity limits access to art and who has the right to speak about it, Eyland argues the opposite. He claims that right-wing attitudes treat art as sacred and make it inaccessible, while the gallery-as-courtroom makes art democratic. Only rational judgments have value in a courtroom; victims' interpretations, because they are irrational, need not be taken seriously.

Eyland claims that metaphorical association is equivalent to superstition, and stands in opposition to dispassionate art viewing. He quotes the comment of victims' rights advocate Walsh, "The same hands that killed thirty-three young boys made these paintings," and comments that "The belief that evil spirits inhabit paintings is consistent with the belief that an art gallery is a sacred space."[7] But in discussion of Gacy's paintings, Baerwaldt's critics never spoke of evil, using instead terms of moral value.[8]

Eyland's courtroom analogy fails on a number of counts, including its assumption that courtroom logic cannot breed right-wing results. Certainly, the experiences of those who have fought censorship of the arts through the courts have revealed just how difficult it is to prove qualitative differences between art and pornography, for example, when empirically they may resemble one another.

Art writers discrediting victims' rights advocates emphasized the superstitious quality of their responses to Gacy's paintings. The word "evil" in these accounts distinguishes the irrationality of Citizens Against Violence from the rationality of art discourse. Introducing the November forum, Noble commented that what interested him was "[the Gacy paintings'] reception by the public as the products or representations of evil, rather than their aesthetic value as art objects." Sharon Alward said,

I thought a lot of the controversy was around whether these things were evil, and how dare we look at them because they may be evil. And they may transmit evil to us, cause [us] to go out and kill people, or think lewd thoughts, or something (Plug In forum 11/27/96).

Actually, none of the members of Citizens Against Violence, or any others of whom we are aware, suggested that the power of the Gacy work was in motivating violence. Instead, they felt its effects upon a more symbolic and metaphorical plane; that these particular art works could, for no discernable purpose, evoke emotions and understandings not only in the families and friends of Gacy's own victims, but also in others who had been on the receiving end of similar crimes.

Baerwaldt's table-turning gesture of framing the *Sun* was his most useful, socially critical response to the issues raised by Gacy's paintings and to the blatant hypocrisy of the *Sun*'s own display of these works. It caustically underlined his curatorial essay's concern with media representation and sensationalizing:

Not ones to let an opportunity for irony slip through it's [sic] fingers, Plug In hung a glass-covered, framed issue of the Winnipeg *Sun* on its gallery wall. Gacy's pictures are prominently displayed in "offset four-color printing" on tabloid newspaper (Buchholz 1996:B4).

Yet Baerwaldt's other expressed concern with discussing how audiences attribute qualities to art with reference to the creator's reputation rather than to any visible qualities of the work itself never seemed to engage his own attention. Given the absence of a forum in which to discuss this point that was of such public concern, his dismissing as superstitious those people who indeed did link Gacy's paintings to Gacy's crimes, and the inclusion of Gacy's work in an otherwise straight (if edgy) art show, made some of Baerwaldt's audience suspect that his own interests did not extend beyond a fascination with "killer chic" (c.f. Achenbach 1994) and the killer's notoriety. Given that Baerwaldt had selected the paintings because of Gacy's notoriety—in other words, for their cultural rather than artistic qualities—the failure to discuss the ways in which materials become culturally meaningful, without dismissing that process as superstitious or irrational, was unfortunate.

But what can these politics of representation reveal about links or disjunctions between art and artists? Edith Regier, who ran a prison art program at Manitoba's Portage Correctional Institute, commented that "pictures are pictures. They are representations. They are not acts . . . No one died from a picture" (1997). True as this is, the evocative power of individual creative expressions is such that the connections between them and their makers are not all that easy to sever. Tracey Walsh was not the only one to suggest that Gacy the murderer produced not only murder, but also paintings; just as Gacy the painter produced not only paintings, but also murder. So did Baerwaldt himself, although he would not see himself as ideologically aligned with Citizens Against Violence in this way.

Although many, like Baerwaldt, might dismiss such beliefs of assumed affinity as mere superstition, the Euro-North American gallery systems also perpetrate such connections of sympathetic magic between artist and work. In the West, the accrual of cultural capital—not to mention exchange value—to art comes in large part through its connection to known, named individuals. Even the most beautifully executed copy of a "Picasso"—note the contagious connection of artist's name and art work—is an imitation, or even a forgery.[9] A painting is what philosopher Nelson Goodman calls "autographic" in that "even the most exact duplication of it does not thereby count as genuine" (1968:113). Indeed, this belief is present in Plug In's solution to showing the work, as its administrator, Bruce Duggan, suggested:

> Those paintings in and of themselves had an enraging talismanic power within this culture. A color copy of them in newspaper didn't . . . none at all. I don't think there was one peep from anybody, ever, except for sour grape art types like me, that how dare the *Sun* show this if it is so bad. Somehow that tiny little step from the object to a reproduction of it drained it of all of that (Duggan 1997).

Euro–North American society extends that type of connection to crime; individuals are the authors of their own crimes, and the correct attribution of illegal activity to its originator is central to the judicial system. Sometimes, lawyers marshal accounts to indicate that the authorship of the crime cannot be entirely autographic, and that external forces, whether Twinkies, PMS, mental illness, or socioeconomic background, reduced personal responsibility. Such arguments become fuel for heated debate around issues of morality. To what extent are individuals truly responsible for their actions, and (how) do sociocultural aspects condition behavior?

These concerns also pertain to the origins of creativity. Artists draw upon sociocultural influences, quote others' achievements, and otherwise dislocate their personal accountability from their works. The distinctions between maker and product, in the contexts of both art and crime, are treated as culturally developed rather than naturally immanent. Yet art works are also personal, individual products. The tension between these aspects is often the source of artistic inspiration and production.[10]

Indeed, the character of the links between Gacy's art and his criminality are the fulcrum of concerns about its display. The issues are complex. Edith Regier commented:

> I don't think that if you draw an image of violence that means you are going to go out and do that. In fact, maybe just the opposite. Maybe you resolve the issues triggering the image enough that maybe you don't have to do it (1997).

Yet assumptions of a direct reflection of personality, character, or identity in art are by no means limited to consideration of Gacy paintings, as Sharron Zenith Corne shows:

> In the early 1970s, I started to do this work which amazed me. . . . [The paintings] were quite sexual and they made me uncomfortable, in a way. At that time there were very few sexual images of women. And I just stored them in a closet . . . until somebody asked if I would show them in their gallery, and I did it reluctantly . . . I think people were shocked by them, and they would not come up to me and say anything that was very revealing. . . . But one of the women who was in my building went up to my brother-in-law and said, "Isn't it good that Sharron can express her sexuality in her work and doesn't have to go around screwing people?" (1997).

Gacy himself asserted in a statement attached to the backs of his paintings: "There is [sic] grave doubts whether he, John Gacy received affective [sic] assistance of counsel, and a fair trial, or as to whether he killed those victims" (c.1990).[11] Tellingly, he never dissociated himself from his art, although his statement is in the third person: "His art seems to imitate parts—but not all—of his life. Generally the darkest corners are eerily absent, lost in riotous colors and often whimsical themes" (ibid.). As Zenith Corne suggested:

> I thought "Perhaps I'm putting it onto it because I knew who did it. Would I feel the same way if I had just come onto the painting at a gallery, not knowing anything about the person who had done this?" . . . Once you know who did them, it is impossible to just dismiss that and look at them in a more distant way (1997).

Yet it seems that just such a process is implied in the *Sun*'s reproduction of the Gacys. Note that the dislocation of image and artist in the *Sun* copy was sufficient for Michael Olito to object to the works' "removal," and for Szu Burgess to comment:

> Initially, I was disappointed, in a way, that the real work wasn't there. . . .
> But I guess because my subject matter is always drawing upon popular cul-
> ture and media propaganda and exploitation and sexploitation . . . that it
> worked better that way, that it was a media representation of this work
> with the text and the skewed interpretation (Burgess 1997).

Citizens Against Violence's Beverley Frey, on the other hand, commented, "I figure they should never have in the *Sun*" (1997). Even
within Euro-North American culture, the purported non-display of the
Gacys is debatable.

BANAL CIRCUS PAINTINGS?
LET'S TALK ABOUT ART

> You can bet that if your great-aunt Maude had turned out the kind of paint-
> ings John Wayne Gacy did, they would not be going on display at the Plug In
> Gallery. Their paint-by-number starkness and stiff, awkward renderings of the
> human form, reminiscent of a high school art class's work, would have rele-
> gated them to yard sale status next to the Elvises done on black velvet. That
> is, unless your great-aunt Maude was a serial killer. Because that's the only rea-
> son Plug In is interested in Gacy's work (Lakritz 1996:5).

For all the focus and attention upon the Gacy works, particularly with respect to their presence and/or absence, it is surprising how little the works'
visuality was addressed. The possibility that form and content can be analyzed in non-artistic terms, or that lack of artistic merit need not affect cultural readings of artifacts (see, for instance, Critical Art Ensemble 1995,
Clifford 1988), remained unaddressed. Arts commentators criticized the
tabloid media for failing to talk about the images themselves or to judge
them on their artistic merit, but the arts community itself never held such a
discussion. They, too, participated in the abrupt dismissal of the paintings'
visuality. With few exceptions, they described the Gacy paintings as banal,
innocuous, poorly rendered, and insignificant. Art critic and writer Robert
Enright commented,

> But even if they were presented, the inexplicable thing for me was the
> paintings were so bad. They couldn't invest [them] with anything evil. . . .
> I tend to, given the tradition I was raised in, give some belief to the notion
> of the aura around an object, so I'm not dismissing that, but these were
> not the objects to invest any argument about evil objects on . . . (Plug In
> forum 11/27/96).[12]

Baerwaldt described the paintings as "technically inept," adding that:

> On first reading certainly they would be so-called bad paintings. You
> might see something sinister or weirdly quirky in them. If [you] looked at
> them—one in particular—and saw the little clown mask that's incorpo-
> rated into one of the portraits, [you] would say, gee, that's funny. It sort of
> looks like something Gacy painted a great deal of the time . . . (*As It Hap-
> pens* 9/17/96).

Baerwaldt argues that the imagery was irrelevant to *The Moral Imagination*. His interest laid, rather, in how Gacy's notoriety triggered expectations in the audience.[13] Such a stance makes it difficult to interpret the
description based on the *Sun* reproductions given by Stephen Phelps, curator of Main/Access Gallery:

> One recoils from the predatory imagery: a hellish apparition hovers men-
> acingly into view astride a merry-go-round horse; a bewigged transvestite
> in a death's head muscle shirt hunkers impassively over some object, the
> painted fingers of an outsized hand splayed like the talons of an eagle
> (1996:A11).

Such descriptions might have provided a starting point for discussion.
What, if anything, does artistic competence have to do with creating a powerful image? What arrangement of paint results in a predatory image? What
prior knowledge, either of art as image or Gacy as predator, or both, is invoked in the viewer? Yet for all the suggestive questions raised around this
very issue, no one, including Baerwaldt, chose to examine it through reference to the works themselves.

Increasingly, artists and curators rely more on art writing than on their
own experience to understand what they are seeing. Such a complaint is not
new amongst artists and art instructors. They attribute this state of affairs to
the professionalization of art training in universities following WW II that
privileged intellectual and academic understandings of art movements while
downplaying the act of seeing. Critic Martha Rosler attends to the possible
consequences of this direction:

> Rhetorical turns common in the art world may seem cryptic, incompre-
> hensible, or insulting to the general audience, or their wider import may
> simply be inaccessible. The invisibility of the message can be ignored by
> art world institutional types, but when community people misread the
> work, the art world must take notice (Rosler 1994:60).

Plug In Gallery and the art practices that inform it grew out of trends in the 1960s. Artists sought liberation from market forces, to make art that could not be commodified, by virtue of size, fixedness, or ephemerality. Artist-run centers, separate from mainstream institutions and commercial galleries, emerged along with the idea of alternative spaces and countercultures. Yet the 1980s was a decade of unprecedented commercialization, and as Martha Rosler (1994) argues, art production increasingly became a branch of the entertainment industry. Rather than resisting the mainstream, segments of the art world mimicked it as a form of sympathetic magic intended to attract some of its wealth to them. That Plug In is not a commercial gallery does not alter the fact that in this case it participated in the logic of capitalist promotion.

Stripped of obfuscating rhetoric, Baerwaldt's gesture to show Gacy's paintings relied on shock value, playing on the killer's celebrity status to attract attention. Baerwaldt claims he intended to comment upon that notoriety. But given that the show as a whole lacked strong curatorial framing, such a "commenting upon" seemed to rest on shaky foundations. Attempting to show Gacy's work was not a gesture of intellectual freedom, but rather an example of art's cooptation into the commodity form. The curatorial action of incorporating Gacy's paintings differs little, if at all, from the *Sun*'s running photos of the paintings to sell papers. The defense of Baerwaldt's curatorial actions in the name of freedom of expression, even if politically expedient, contributed to the mystification of vision in art. Analyses of the artifacts, the curatorial intent, and the success of the work were given short shrift.

SYMBOLIC EFFICACY/CREATIVE ALTERNATIVES

Critic Garth Buchholz, in his review of *The Moral Imagination* in the Winnipeg *Free Press,* describes Tony Tasset's "Squib" in terms obviously influenced by issues that others related primarily to the Gacys:

> . . . a kind of snuff video in which a man stands almost still with his back against the wall. . . . Suddenly, and highly realistically, a gunshot explodes in a flash, the man is thrown back against the wall and his "blood" appears . . . on his chest and the wall. After watching this disturbing video loop play again and again, even an art critic who has seen it all has to marvel at the detachment with which some artists can deal with moral issues. Those who have lost a family member through violent crime may find that they do not need to attend this exhibit to "get the message" (Buchholz 1996:B4).

This perception begs a question not publicly addressed by Plug In, but of great concern to some of *The Moral Imagination*'s audience. How can a representation of violence, in whatever form, be socially or artistically critical? Baerwaldt's curatorial silence on this issue failed to illuminate it, but Edith Regier's argument accounts for both censorship concerns and social responsibility:

> If we don't look at [violence] then for one thing we will continue to be naive about it. I mean, how do people get away with this stuff? . . . I don't know if you necessarily have to glamorize violence just to look at it. To single out and to focus so much attention on the Gacy paintings did work to glamorize him and this acts to separate his images and actions from all the other incidents which happen, which are so similar, and doesn't really help towards understanding how a human being develops into someone . . . who is a sociopath, who lies, and is not in touch with his or her actions. I mean, I do think that is what those paintings look like (1997).

And, although the *Sun* pilloried Baerwaldt's plan to show the Gacy paintings, clearly the tabloid's own hasty display failed to address their own purported concerns of insensitivity and exploitation. Grant Guy said,

> I think that it was more interesting that the reproductions were there as opposed to the real thing. . . . Stealing from the Winnipeg *Sun* changed the context one step further. I found it interesting that, at the end, the Winnipeg *Sun* had the smoking gun, and not Plug In (1997).

But couldn't Plug In have done better than the *Sun?* As Hans Haacke suggests, "The problem is not only to say something, to take a position, but also to create a productive provocation" (Bourdieu and Haacke 1994:20–21). Far from upsetting the mainstream establishment, Baerwaldt's plan to show the Gacys—perhaps unintentionally—provoked an already marginalized group, and his justifications in the name of freedom of expression failed to convince them.

In essence, Plug In reproduced the tabloid discourse, facing the same old choices identified by art critic Mary Devereaux:

> Either we embrace the political character of art and risk subjecting art and artists to political interference, or we protect art and its makers from political interference by insisting upon their "autonomy," but at the cost of denying the political character of art and its broader connection with life (1993:207).

We are all affected by crimes, by watching them, reading about them, knowing that they exist. They change us, our subjectivity, our sense of what is human, what is evil, and what is moral. It is the larger effect of violent crimes that can be addressed creatively by gallery art. The planned Gacy exhibit provoked response by directing the public to associate the paintings with Gacy's murderous acts. Alternatively, it could have decentered Gacy's voice and his weirdness, to complicate the issue of violence rather than just shoving it in peoples' faces; to give the victims a living voice; and to allow others to express healing, fear, and anger. But in the absence of a larger interpretive frame through which reactions could be analyzed and—perhaps even more importantly—understood, the potential for suffering to be relived without being redeemed was kept alive. As Victor Turner might have noted, the ritual drama of art-viewing would have failed to link bodily experience to moral value, since Baerwaldt's "Moral Imagination" itself appeared curiously amoral. The Gacy paintings, together with the careless way in which their showing was proposed and defended, perpetuated the marginalization of victims' concerns and demonstrated a curious blindness to the complex power of art.

Baerwaldt's one-liner solution of framing the *Sun* remained his only productively critical, implicitly politicized response joining the two politics of representation around this issue. But to actually challenge hegemony would have necessitated the creation of "communities of interest," as advocated by artist Martha Rosler—between victims of violence, artists and curators (commonly subject to tabloid exploitation); between the arts community and medical workers (equally threatened by downsizing and restructuring); and so on. Here, perhaps, Baerwaldt's own moral imagination failed.

ACKNOWLEDGEMENTS

We would like to express our appreciation to Sharon Condie for invaluable research, and to Sarah Koch-Scholte and Cynthia Thoroski for transcription. Pauline Greenhill thanks the University of Winnipeg for a Discretionary Grant. Anne Brydon expresses her appreciation to the Agnes Cole Dark Fund, Faculty of Social Science, University of Western Ontario, for receipt of a research grant. We appreciate useful criticism from Joyce Mason, Leslie Korrick, and Stephanie Kane, which helped us to hone the analysis. We especially thank Plug In Gallery's board and personnel, particularly Wayne Baerwaldt and Bruce Duggan, for their friendly cooperation.

NOTES

1. While in prison awaiting his May 1994 execution in Illinois, after his 1980 conviction for the sexual torture and slayings of 33 young men and boys between 1972 and 1978, Gacy produced some hundreds of paintings and sold

them to pay for his defense. He often depicted circuses and clowns. (Before his arrest, Gacy, as Pogo the Clown, gave charity performances for children, where sometimes he chose his victims.) More information on Gacy, his crimes, and his case can be found in Wilkinson (1994).

2. Certainly this fact in itself could provide fertile ground for interpretation of the paintings, to think about the relationship between a self-loathing serial killer acting out his internalized homophobia and a cultural producer's consumption of the perverse. Yet Baerwaldt never revealed the identity of the paintings' owner, referring instead to an "anonymous L.A. collector." We don't know why Baerwaldt kept Meadmore's identity as owner of the paintings confidential. It was not at Meadmore's own request; he wants to sell the paintings.

3. Interviewing Baerwaldt, Brydon asked him about the content of the paintings. At first he said it was beside the point, then he added, "How do we know what is in these paintings? Maybe it is a torso. Maybe it is a badly rendered footstool."

4. *The Moral Imagination* included the following artists: from Winnipeg, Eleanor Bond, Szu Burgess, Sharron Zenith Corne, Grant Guy, Patrick Hartnett, Jake Kosiuk, Bonnie Marin, Al Rushton, Harry Symons, and Evan Tapper; from the United States, William S. Burroughs, Iñigo Manglano-Ovalle, Kevin Mutch, Nam June Paik, Roy Pardi, Bill Smith, Tony Tasset, and Barbara Wiesen; from Chile, Juan Davila; from France, Pierre Molinier; from Austria, Hermann Nitsch; and from Brazil, Carlos Zéfiro.

5. See, for example, the discussion of Benetton ads in Giroux (1994).

6. As suggested, for example, in Critical Art Ensemble (1995).

7. Walsh's statement mirrors what a defender of "killer chic" wrote in the fanzine *Answer Me!:* "'Whatever the technical merits of a finished piece are, it was drawn by the same hand which has taken lives. Consciously or not, a killer gives vent to the clotted backwaters of his harrowed cranium'" (Washington Post 2/13/94: F7).

8. The idea of evil comes from a *New Yorker* article by John Updike (1996) that Baerwaldt had read around the time he selected the Gacy paintings for the show. Updike does not distinguish evil's wickedness and religiosity from morality's concern with social right and wrong. Such a distinction struck us as relevant when considering how to engage discussion about *The Moral Imagination.* When asked if his critics thought showing Gacy's paintings was immoral or evil, Baerwaldt could not distinguish between the two terms.

9. This is a curiosity that Jorge Luis Borges engages in his essay "Pierre Menard, Author of the *Quixote*" (1962).

10. Foucault's "What Is An Author?" (1975) concerns such issues in other contexts. Indeed, anthropological investigation problematizes to what degree

individuals are free and have agency, and to what degree their actions are socially constructed and constrained.

11. Some people have been convinced by Gacy's arguments of his innocence and conclude that some other individual must have buried the 27 bodies in Gacy's basement.

12. Enright's reference to the aura refers to Walter Benjamin's discussion of mechanical reproduction's impact on the singularity of the artistic production. Modernist art critiques the power of the aura in a move similar to critiques of ideology (cf. Buck-Morss 1989).

13. This fascination with the artist as icon carries on in his recent curatorial project that includes the work of Yoko Ono.

BIBLIOGRAPHY

Achenbach, Joel
1994 Killer Chic: It's Ugly, But Is It Art? Washington Post, Washington, D.C., February 13: F7.
Baerwaldt, Wayne
1996 The Moral Imagination. The Plug In Harold, October: 1–10.
Beardsley, John
1995 Gardens of Revelation: Environments by Visionary Artists. New York: Abbeville Press.
Becker, Carol
1994 Introduction: Presenting the Problem. *In* The Subversive Imagination: Artists, Society, and Social Responsibility. Carol Becker, ed. Pp. xi-xx. New York and London: Routledge.
Bhabha, Homi K.
1994 The Location of Culture. London and New York: Routledge.
Borges, Jorge Luis
1962 Labyrinths. New York: New Directions.
Bourdieu, Pierre and Hans Haacke
1995 Free Exchange. Stanford: Stanford University Press.
Bray, Allison
1996 Crossing the Line: Controversial Art Draws Fire. Winnipeg Free Press, October 18: C1.
Buchholz, Garth
1996 If, at First, You Don't Succeed Try . . . The Free Press, Winnipeg, November 2: B4.
Buck-Morss, Susan
1989 The Dialectics of Seeing: Walter Benjamin and the Arcades Project. Cambridge, MA.: MIT Press.
Burgess, Szu
1997 Interview, July 12. PG97–17, 18, 19.

CBC One
1996   As It Happens. September 17.
Clifford, James
1988   The Predicament of Culture. Cambridge, MA: Harvard University Press.
Critical Art Ensemble
1995   Human Sacrifice in Rational Economy. Public 11: 125–134.
Devereaux, Mary
1993   Protected Space: Politics, Censorship, and the Arts. The Journal of Aesthetics and Art Criticism 51 (2): 207–215.
Duggan, Bruce
1997   Interview, January 24. PG97–1, 2.
Eyland, Cliff
1997   On Temples, Trials and Taste. Mix 22(3): 61.
Foucault, Michel
1975   What Is An Author? Partisan Review 43: 603–614.
Frey, Beverley
1997   Interview, February 22. PG97–3, 4.
Gacy, John Wayne
c. 1990 About the Artist. Ms. attached to painting.
Giroux, Henry A.
1994   Benetton's "World Without Borders": Buying Social Change. In The Subversive Imagination: Artists, Society, and Social Responsibility. Carol Becker, ed. Pp. 187–207. New York and London: Routledge.
Goodman, Nelson
1968 The Languages of Art: An Approach to a Theory of Symbols. Indianapolis: Bobbs-Merrill.
Guy, Grant
1997 Interview, July 8. PG97–13, 14.
Lakritz, Naomi
1996   Gallery Exploits Shock Value. The Winnipeg Sun, September 18: 5.
Laser, Sig
1997   Interview, February 27. PG97–5,6.
Maizels, John
1995   Raw Creation: Outsider Art and Beyond. London: Phaedon Press.
Mitchell, Scott
1997   The Moral Imagination: Some Background. MIX 22(3): 62–63.
Noble, Richard
1997   Art, Morality and Mr. Gacy's Paintings. Border Crossings 16(1): 14–19.
Olito, Michael
1997   Interview, July 10. PG97–15, 16.
Phelps, Stephen
1996   Some Art Represents So Much Evil As To Be Unchangeable: Gacy Plan a Bombshell. Winnipeg Free Press, October 1: A11.

Regier, Edith
1997   Interview, April 28, PG97–9.
Rosler, Martha
1994   Place, Position, Power, Politics. *In* The Subversive Imagination: Artists, Society, and Social Responsibility. Carol Becker, ed. Pp. 55–76. New York and London: Routledge.
St. Germain, Pat
1996a Local Gallery To Exhibit Killer's Death-row Art: Gacy Paintings Anger Victims of Violence. The Winnipeg Sun, September 17: 3.
1996b Tax Bucks for Sick Art Irks Public. The Winnipeg Sun, September 18: 3.
1996c Death-row Daubings Unplugged. The Winnipeg Sun, September 26: 5.
Updike, John
1996   Elusive Evil: An Idea Whose Time Keeps Coming. The New Yorker, July 22: 62–70.
Wilkinson, Alex
1994   Conversations with a Killer. The New Yorker, April 18: 58–76.
Zenith Corne, Sharron
1997   Interview, April 27. PG97–7: 8.
Zolberg, Vera L. and Joni Maya Cherbo, eds.
1997   Outsider Art: Contesting Boundaries in Contemporary Culture. Cambridge: Cambridge University Press.

CHAPTER 7

CRIMINAL INSTABILITIES
Narrative Interruptions and the Politics of Criminality

JoAnn Martin

I EXAMINE HERE THE TECHNOLOGIES OF POWER at work when political figures are constructed as criminals in regimes in which, however thin, the veneer of political freedom matters. In such situations, the establishment of the "truth" of charges of criminality is central. I focus on a historic example, the construction of Emiliano Zapata, leader of the southern forces of the Mexican revolution of 1910, as a criminal.[1] My discussion takes place against the backdrop of my ethnographic research in Buena Vista, a town in Morelos, Mexico, where Zapata is today considered a local hero, and where residents have experienced the Mexican state's use of criminal charges against political activists.

Shortly after his assassination at the hands of government troops in 1919, Zapata was reconstituted as a national hero, first as a champion of the causes of poor peasants and later as a kind and paternalistic figure who protected the interests of both rich and poor alike (O'Malley 1986). Zapata is now so entrenched as a national hero that attempts to rewrite Mexican history books in the 1990s to provide what then president Carlos Salinas de Gortari called a more accurate portrayal of the revolution caused a great uproar in Mexico.

Understanding the processes of social transformation that underlie the instability of criminal and revolutionary political identities entails considering how criminal identities are momentarily stabilized within revolutionary settings. Reporters covering the Mexican revolution did not all concur that Zapata was a criminal. Some argued that Zapata's followers were barbarians, but that Zapata himself was motivated by noble principles. Others rejected the criminal characterization altogether, asserting that the government troops were the true criminals. But attempts to complicate the image of Zapata floundered under the weight of interpretive practices that reveal the place of reporters in governmentality.

Michel Foucault uses the notion of governmentality to refer to the art of governance in which power is shaped around "the life-conduct of an ethically free subject" (Gordon 1991:5). As a correlative of ethically free subjects, new complex technologies of power come into being for producing truth. Emphasis is placed on documenting, analyzing, evaluating, and diagnosing the state of the population—its physical and mental health, its biological reproduction, its labor capacity, and its consumption needs so that relations between people and things can be managed through scientific knowledge (Foucault 1991:93). Governmentality takes place, therefore, at the juncture between self-governance, fostered by the deployment of discourses of truth, and the art of governance understood as the management of a delicate set of relationships between persons and things. As Foucault points out in his conclusion to *The History of Sexuality*, governance turns on the protection and administration of life in all its rich detail (1990:135–159).

If the art of governance entails striking the perfect balance between the intervention of the state and self-governance, crime threatens this delicate equilibrium. Crime suggests a failure of those technologies of power aimed at producing docile bodies; and yet, if governmentality is to prevail, the crime and the criminal must be subjected to technologies of power consistent with the reasoned management of society. There must be techniques for producing "truth" regarding criminals so that they can be reconstituted as manageable cases, taking their place among other deviant, albeit manageable subjects, such as the insane, the perverse, women, and children. Finally, in a society of free subjects, the production of truth about the criminal must be deployed widely, even becoming the subject of debate. Criminality is compatible with governmentality only when the criminal is shown to be a manageable subject, an object of interest and investigation, rather than a terrifying figure.

If criminality poses challenges to governmentality generally, the establishment of criminality in a revolutionary context raises special problems. While all crimes may inspire interest in the criminal subject, the terror of revolutionary violence announces the possible failure of governmentality by raising the specter that things have become unmanageable. Moreover, revolutionary violence challenges governmentality at its core by attacking the idea of neutral management of the relationship between people and things.

The dilemma for governmentality in the face of revolutionary violence is the development of techniques and methods for the management of terror. If under ordinary circumstances the response to criminality is to contain the terror inspired by the criminal, in a revolutionary context governmentality depends upon both provoking and managing terror. If in day-to-day criminality, criminal acts are eclipsed by the emergence of the criminal subject, this is not true in the establishment of criminality in a revolutionary context. When criminal acts are carried out in the name of revolutionary agendas, the criminal must always share the stage with his or her acts. Indeed, the "criminality" of such acts is established at the interface of the criminal subject and the toll taken on specific lives in progress.

Mexican reporters helped solidify Zapata's reputation as a criminal by marking Zapatista attacks as a threat to a "life in progress." The portrayal of such "lives in progress" highlighted the Zapatistas, not as threatening a particular government or policies, but as a danger to governmentality in general. Elite life was shown as the intersection of self-governance and the proper management of society that suddenly and violently was disrupted by Zapatista attacks. If train schedules were disrupted and hotels unsafe, all aspects of elite life were placed in jeopardy. Family relations, the proper form of gender roles, and the smooth management of estates depended upon governmentality. Reporters' descriptions effectively captured the mutual accommodation between those who govern themselves and the proper administration of territory that enables the freedom of self-governance.

A key rhetorical device used in reports of the revolution was the representation of Zapatista attacks as an interruption of everyday routines governed by norms of civility. Reporters' narratives described in detail what elites were doing when Zapatista attacks took place. Not surprisingly, it is precisely this sense of criminality as an attack on a life in progress that is missing from reports of government attacks on villages believed to support the Zapatistas. Even those sympathetic to the revolutionary forces seemed incapable of presenting the civility of peasant life. The inability of reporters to capture the life in progress of peasants in a fashion analogous to their treatment of elite culture speaks to the intersection of governmentality and hegemony in the early 1900s. Reporters' observations highlight elites as having cultivated the disciplinary routines of a free society, against which Zapatista criminality emerges.

If reporters easily established Zapatista criminality, the connections to Zapata remained elusive. Peasants claimed that government troops were dressing as Zapatistas to discredit the revolutionary movement, and the loose coordination of the Zapatista movement made it easy for bandits to invoke the revolution while carrying out criminal acts. It fell to reporters to establish the connection by representing Zapata as a criminal type, drawing on turn-of-the-century criminal anthropology. Debates then raging in criminal anthropology were brought to bear on first-hand examination of Zapata's character carried out by investigative reporters. Some maintained that Zapata was a strange human/animal hybrid capable of enduring a harsh environment, while others associated him with the criminally insane. Still other reporters drew on Zapata's own words to produce laudatory and humanistic descriptions of the toll fighting the revolution was taking on his body.

EMILIANO ZAPATA: THE CENTRAL CHARACTER

To peasants in the town of Buena Vista, Emiliano Zapata is affectionately remembered as a *local* hero who shaped the nation and spoke for the just demands of the poor. Claiming Zapata as their own, peasants seek indications in their personalities and local stories of genealogical links to Zapata, or at least to Zapatista fighters. Zapata's words, "land to those who till the soil," hung on a banner in the office of communal land where I first carried out my research. Older peasants proudly shared collections of Zapatista memorabilia with me, and younger peasants gave me books to read on Zapata. Everyone told me stories to introduce me to the true Zapata, not the one the government had co-opted.[2]

Those old enough to remember the revolution recalled the events for which Zapata became known as a criminal. Either because of his reconstitution as a national hero or because they never viewed his acts as criminal, they talked of Zapata's acts as inevitable aspects of revolutionary war. Buena Vistans lamented that land documents critical to protecting the town's boundaries were destroyed when Zapatista fighters burned the *municipio.* Older women who were young girls at the time described living in fear of being kidnapped by Zapatista fighters who would take them into the mountains, rape them, and require them to travel with and cook for them. They noted that government troops often did the same. But, with the passage of time, even these fears were reconstituted. One women in her nineties insisted that she had been kidnapped by a Zapatista who raped her and kept her with him until he was finally killed. Showing me a picture of him in his Zapatista uniform, she proudly boasted that he would say, "Marta, I want to keep you in my pocket so I can take you everywhere I go." Others in town claimed Marta had been captured not by a Zapatista, but by bandits who roamed the area after the revolution. The categories of *criminal* and *political* remain unstable.

Figure 7.1   Zapata was represented as a frightening hybrid combining human insanity and paranoia with animal-like strength and endurance. Reproduced from the collections of the Library of Congress.

Emiliano Zapata was born in the late 1870s into a Morelos family that would have been considered better-off: He farmed land inherited from his father and sharecropped additional hectares from the local *hacienda* (plantation) (Womack 1968:6). An accomplished horseman, he worked as a trainer for many *hacendados* (planters), although his experience with "polite society" is said to have left him with great disdain for the differences in wealth between rich and poor (ibid.). Zapata's rejection throughout his life of opportunities to climb the ladder of elite society firmly marked him as a man of the people. As Womack (1968:7) describes, "If he [Zapata] dandied up on holiday and trotted around the village and into the nearby town of Villa de Ayala on a silver-saddled horse, the people never questioned that he was still one of them."

Zapata's struggle for land and liberty represented both continuity with a long history of land conflicts in the state of Morelos and the outcome of forces that amplified the impact of land shortages in Morelos in the early 1900s. Under Porfirio Díaz, president of Mexico between 1877 and 1880, and again from 1884 to 1910, Mexico embraced foreign capital and expanded its rail system. Díaz's policies were supported by cabinet members and intellectuals influenced by Auguste Comte's postivist philosophy (see Comte 2001). Known as *científicos,* these intellectuals attempted to use statistics and the science of sociology to transform Mexico into a modern nation (Miller 1985:266). Their presumption, reinforced by Darwin's notion of survival of the fittest, was that Mexico's indigenous population and its tradition were destined to disappear. Accordingly, the científicos provided the intellectual justification for industrial expansion at any costs; and, under Díaz, the state supported capitalist development by welcoming foreign capital, establishing rail connections throughout the nation, and modernizing the mining industry.

Planters in the state of Morelos responded to the ideals of order and progress professed by the científicos by importing heavy machinery that enabled them to increase sugar production (Wolf 1969; Womack 1968:15,49). With the promise of higher profits, hacendados began to expand their holdings—sometimes through purchases but often through illegal takeovers of peasant and village property. President Díaz supported expansions of the haciendas with legislative reforms that legalized the land takeovers (Wolf 1969:16). By 1908, Morelos's plantations ranked the state third in production among sugar-producing regions of the world (Womack 1968:49). Under the weight of Díaz's policies and the incentives for sugar production, many Morelos communities disappeared as haciendas absorbed the surrounding communities; others lost great extensions of communal land. These land pressures combined with the erosion of community autonomy to prompt many Morelos peasants to fight for their lands through legal and political means. Eventually many joined the revolution.

Between 1909–1910, Morelos villagers throughout the state took complaints about local hacendados to the courts and petitioned Díaz himself to rectify disputes. Zapata emerged as a local hero when he successfully resolved one such dispute between his hometown of Anencuilco and the hacienda hospital. By the winter of 1910–1911, Zapata was acting as an authority, resolving land disputes in the wealthiest sugar producing regions of Morelos (Womack 1968:75). In so doing, he established the organizational foundation for Morelos's participation in the Mexican Revolution.

At the same time that Morelos villagers contested land takeovers, forces from around the country were converging around a plan to dislodge Díaz from the presidency. At issue was not so much Díaz's economic policies, for which he maintained considerable support among the elite, but the absence of political freedom (Miller 1985:281). Díaz himself encouraged some of the discontent in 1908 when he promised to retire in 1910 and invited the formation of an opposition party (Bazant 1977:122). Franciso Madero, a wealthy, northern landowner took Díaz at his word and tried to form an opposition movement, but Díaz retaliated by oppressing the movement. When Díaz again won re-election in 1910, Madero was jailed, although he later escaped and proclaimed the revolution in the name of democracy. Hoping that Madero would protect village land, Zapata and the Zapatistas joined forces with the Madero revolt in the spring of 1911 (Womack 1968:80). Although Madero's commitment to land reform and community independence remained questionable, the alliance with Madero helped to legitimize the Zapatistas as participants in a broader revolutionary movement.

ZAPATISTA CRIMINALITY

The elite of Morelos began to see Zapata and the Zapatistas as a threat to the civil order for what they described as his lack of respect for the rights of private property and for his embrace of robbery and pillage over the merits of an honest day's work. As the revolution unfolded, elites used their control over the press to charge Zapata with murder, rape, torture, destruction of civil property, and disruption of transport throughout central Mexico.

In the scholarly literature on Zapata, including Womack's (1968) seminal biography, tales of Zapatista criminality have been treated as an exaggeration encouraged by elites threatened by the revolutionary leader's political agenda (Brunk 1996:332). Samuel Brunk's (1996) article "The Sad Situation of Civilians and Soldier: The Banditry of Zapatismo in the Mexican Revolution," provides the only systematic treatment of criminal acts committed by Zapatista forces. Drawing on Zapata's personal archives and communiques issued by Zapatista headquarters, Brunk shows that the banditry that occurred among Zapatista forces was often directed at peasant villages. He paints a picture of Emiliano Zapata trying in vain to control robbery and pillaging by his forces and of Zapatista chiefs charging one another with banditry. Brunk emphasizes, however, that the problem of banditry emerged after 1915, when many hacendados had left Morelos and villagers and Zapatistas had to compete with one another for food.

In contrast, Brunk characterizes the period between 1911 and 1914 as one in which "the banditry that occurred was generally rather tame and limited: property was appropriated, but there was little serious violence against the hacendados and their allies" (Brunk 1996:337). Yet, during this early period the most detailed and damaging articles about Zapata and his movement appeared in the Mexican press.

Brunk acknowledges that the Zapatistas in their early period (1911–1914) may have infringed upon legal norms, but he suggests these infringements must be placed in the context of the history of land expropriation by hacendados under Díaz. He cites Zapata's December 1911 response to charges of banditry: "One cannot call a person a bandit who, weak and helpless, was despoiled of his property by someone strong and powerful, and now that he cannot tolerate more, makes a superhuman effort to regain control over that which used to pertain to him. The despoiler is the bandit, not the despoiled!" (cited in Brunk 1996:337).

Of course, Zapatistas did more than appropriate property; they attacked the very fabric of governmentality. They released prisoners, helped themselves to food and liquor, destroyed municipal archives, sacked and burned businesses, and targeted the properties of Spaniards as representatives of foreign oppression (ibid.:335). Zapata himself seems to have condoned these activities and was present at many of them, at least until he realized that within the wider society these acts caused legitimacy problems for his movement (ibid.). Ultimately, however, Brunk suggests, "This behavior fits our definition of banditry, but it was also quite clearly class conflict and had an underlying revolutionary rationale" (Brunk 1996:335).

Clearly, Brunk's view of early Zapatismo leans on insights afforded by the passage of time, and he and others are correct to argue that "criminal" acts of this period are a window into revolutionary strategies. His analysis supports Zapata's later reconstruction as a national revolutionary hero, but his article does not address how interpretive practices of elites constructed Zapatista criminality in the first place.

Although, in the final analysis, Mexican elite constructions of Zapatista criminality may well have been a reflex of their property interests, narratives that appeared in the press at the time united elite sentiments around a far more compelling story: Zapatista "criminality" threatened the ongoing "lifestyle of the elite" and their place in the political economy of Mexico. Worse, Zapata and his men might gain the political legitimacy that could lead him to be governor. Indeed, the specter of Zapata as governor emerged in newspaper editorials, making the need to prove the revolutionary leader's criminality even more pressing. To establish Zapata's criminality, contrary to legal strictures that emphasized criminal acts, journalists of the period drew on turn-of-the-century criminal anthropology as indices of Zapata's underlying character abnormalities. Zapata was represented as a frightening hybrid combining human insanity and paranoia with animal-like strength and endurance.

INTERROGATING TERROR

Newspaper accounts invariably began with a promise not to exaggerate but to relate only the facts as they had occurred, implying that while the terrors of war made it impossible for other experts to engage their analytical skills, reporters could be objective, scientific observers. Their job became the detailed and systematic portrayal of Zapatista attacks. That journalists contextualized these attacks around the routines of elite life suggested an easy familiarity with elite culture. The following excerpts about a train attack appeared in the newspaper *La Prensa* on March 31, 1912:

> The journey was proceeding happily, although among the passengers there reigned a fear of assault which the captain of the escort tried to calm with particular attention to the women. A few hours before the passenger train had departed a cargo train had derailed either by accident or by attack of the Zapatistas . . . the passenger train was slowing down to avoid collision when a direct hit by a pullet that issued from between some nearby rocks took the life of the train engineer Villegas who had barely enough time to halt the train before falling dead. . . . As soon as the train halted a rain of bullets began to fall causing panic among the passengers who hid under their seats. The escort responded bravely firing from the windows of their car. As the combat continued Captain Melo tried to move the battle away from the train so as to avoid deaths among the passengers. He descended from the train and shouted to his men to do so. No one responded and in indignation he climbed back into the car only to find it crammed with cadavers—not a single soldier remained standing . . .
>
> By now the Zapatistas had left their hiding places and headed threateningly toward the train. The assailants mounted the train forcing the passengers to pass between the bodies of the dead and wounded to descend the train . . . The death of Señor Juan Veraza at the hands of bandits is now confirmed. As the victim kneeled begging for his life the bandits killed him with two blows from the butt of their rifles. Also confirmed is the assassination of young Francisco Napoles whose body was viewed by our reporter. This young high school student was traveling with his brother to Cuernavaca to enjoy a vacation with his family . . . [3]

On April 1, 1912, *El Diario del Hogar,* a newspaper generally more sympathetic to the Zapatistas, carried an interview with the brother of young Francisco Napoles:

Reporter: Please tell us how your brother died.

Answer: Once the Zapatistas had fired upon the train, killing the heroic federal soldiers leaving only the injured Captain still standing, the Zapatistas mounted the train. They discovered the Captain in the process of changing his clothes and were going to kill him right there. My brother tried to push away one of the rifles that was pointed directly at the Captain's chest. The bullet hit my brother. He fell immediately and breathed a final long breath. I tried to revive him asking him where he was hit but he never answered me . . .

Such descriptions of revolutionary violence seasoned daily reporting on the Mexican Revolution and constituted the Zapatista rebellion as an especially barbaric attack on civilization. The power of these stories to evoke terror leaned upon the unpredictability of violence and the representation of lives it destroyed. Narratives of purported Zapatista attacks—few newspapers bothered to determine who the attackers really were—repeatedly began with images of the routines of everyday life interrupted by violent attacks that would change the course of victims' lives forever: the brother traveling home to enjoy vacation, the peaceful first-class hotel disrupted by gunshots from Emiliano Zapata's brother, the wife of a Spaniard threatened in her hotel room, the hacendado attacked in his home as he tried to carry on the work of managing his estate.[4]

Newspaper reports established the tragic on the value of particular ways of life, not in relation to the legitimacy of the Mexican state which, in the midst of the revolution, was indeed questionable. Reporters' vignettes sharpened the story of violence by invoking human victims as caricatures of normal life: brothers, hardworking students who loved their families and shared their final moments together, males who protected women from fear, and heads of households who defended their property for their heirs. The reports pointed therefore not to the ways Zapatistas threatened state power, but to the threat they posed to details of everyday life that symbolized a sense of well being. As relayed in reports of violence, those on the train spoke for the intersection of what Foucault terms the triad of "sovereignty-discipline-government" (1991:102). The tragedy of violence was cast not then as a disruption of an economic and political system, but as violations of happy and peaceful lives snatched away at the very moment of their elaboration.

In contrast to the contextualization of Zapatista violence as a disruption of everyday elite life, reports of government violence against peasant villagers lacked the sense of valued lives interrupted. This remained true even in papers such as *El Diario del Hogar,* which were largely supportive of Zapatista demands for land.[5] To be sure, these papers did report the fear and terror that consumed village life as villagers endured attacks from government troops. Passionately written reports condemned government actions, but they lacked the sense of everyday routines that contextualized Zapatista attacks on elite culture. The following story, which appeared in *El Diario del Hogar* on November 16, 1911, describes a reporter's visit to the town of Tepoztlan, Morelos:

> Today with more calm, I undertook an investigation of the population and saw that the terror that reigns remains as great as it was during the last day of the siege and combat. To announce the presence of troops they fired a canon shot . . . and everyone prepared for another sacking. The women and children began to scream . . . many ran toward the mountains to find shelter in the forests. Some of those injured have gone to the governor to inform him of their complaints and have been harshly received. . . . The governor told the honorable officials of the municipio that it is the rabble that is committing the pillage. One of the officials, demonstrating his bravery, responded that there is no such rabble in Tepoztlan—all are honest and hard working and therefore the only ones that could be committing these crimes are the painted soldiers [name used for government troops].

The systematic absence of a sense of peasant life in progress suggests the position of reporters, even progressive ones, within the hegemonic structures of Mexican society. While reporters might arrive in a peasant community to cover the aftermath of an attack, the day-to-day lives of these communities remained outside their imaginative frameworks. In the absence of descriptions of individual lives in progress, reports focused on the community as a kind of collective victim. Unlike the descriptions of Zapatista attacks, no individuals emerge in attacks on peasants. Instead, reporters' accounts invoke mass subjects, sometimes divided by gender and age, running and screaming. Missing is the kind of detail that might create the imaginary links between these people fleeing violence and the lives of those reading newspapers. If, as Benedict Anderson (1983) has suggested, the emergence of print media made it possible to "imagine" the nation, slight differences in the way reporters contextualized the violence of the Mexican Revolution created systematic boundaries around the way the nation was imagined. Indeed, if the científicos had believed that peasants were to disappear, newspaper reporters reinforced this perception by failing to capture the everyday routines of peasant life in their reports of violence. The attack on elite Mexican society wrought by the Zapatistas emerged as a threat to all the accoutrements of higher civilizations; those on peasant communities demonstrated only the brutality of government troops and not the worth of the communities themselves.

Images of the intimacy of violence, its penetration into the protected spaces of elite culture, and its violation of elite life as an aesthetic work in progress infuriated readers of Mexican newspapers. These journalistic images reinforced a political agenda designed to delegitimize the Zapatista movement and to assure that Zapata and his men would never acquire formal power in the post-revolutionary government.[6]

ZAPATA AS A CRIMINAL CHARACTER

A central tension that emerges in reports on Zapatista criminality lies in the difficulty of establishing links between Zapata and acts committed by his followers. While some reporters made no effort to discern differences among Zapatistas, others relentlessly asserted that Zapata differed from the many bandits who used the cover of the revolution to rob and steal. Formal criminal charges were never filed against Zapata (Brunk, personal communication), although the English language paper, the *Mexican Herald,* and *El Imparcial,* a newspaper sympathetic to the hacendados, demanded a criminal investigation of Zapata. An editorial in the *Mexican Herald* laid out the charges against Zapata.

> Let us look at the situation: here is a man, who without holding any civil or military position is virtually the arbiter of the destinies of the state of Morelos. By his mere fiat he stops the movement of trains and in other ways interferes with communication becoming by that very act guilty of a felony. He maintains an armed force, subject to his command without legal authorization and for no legitimate purpose, and lays down as a condition for disbanding that force that the government withdraw its regular troops, belonging to the national army . . . what is going to happen if Zapata is willing to lay aside his contumacious attitude? Is he going to be allowed to retire honorably into private life? He against whom criminal charges are pending? Is no inquiry to be made into Zapata's share of responsibility in the cruel murder of Carmen Garcia former secretary of the Jefatura at Cuautla? Are Zapata's lieutenants, some of whom are charged with participation in the murder of Spaniards in the state of Puebla to be allowed to go their way unmolested? . . . At any rate to negotiate with such men, to allow them to treat with the government on a footing almost of equality, would surely be to set a dangerous precedent. Zapata must be brought to book and the principle of authority enforced . . . [7]

The death of Carmen Garcia, referred to as the political head of the town of Cuautla, is the only case I found in which the newspapers directly link Zapata to an assassination. In May 1911, Zapatista forces, under the command of Zapata, moved against Cuautla, which was protected by the Fifth Regiment (Brunk 1995:37). After a costly battle with losses on both sides, federal troops were forced to abandon the town to the Zapatistas (ibid.). The brutality of the battle provided newspaper reporters with considerable grist for elaborating Zapata's brutality. According to *El Imparcial*, Carmen Garcia, a simple civil servant, was brought in front of Zapata. After hearing the charges against him, Zapata declared, "This one I will shoot personally," and fired three shots. According to the paper, with Garcia barely alive and still writhing in pain, the Zapatistas tossed his body into a well and finished him off with rocks.[8]

The *Mexican Herald* provided a motive for Zapata's brutality against Garcia. They claimed that prior to the revolution, Zapata, in legal trouble, had been forced to serve in the 29th Infantry. Carmen Garcia, at the time secretary to the political head of Cuautla, made out the papers that consigned Zapata to the infantry. The newspaper concluded that Zapata's desire to personally execute Garcia must have been motivated by revenge, but revenge against a mere civil servant carrying out a task under orders of his superior![9]

In the description of Zapata's attack on Garcia, the *Mexican Herald,* acting as prosecutor, supplies a motive for the crime created around Zapata's misunderstanding of the routines of everyday life that represent governmentality. In the newspaper account, Zapata failed to understand the position of a simple civil servant, Garcia, who had a job to do and a boss whose orders he had to follow. That the Zapatistas reportedly freed prisoners, burned down muncipal buildings, robbed stores, and saved their special revenge for the Spaniards only further buttressed the notion of Zapatista criminality.[10]

After a detailed description of Zapatista atrocities, the newspaper *El Imparcial* attempted to enumerate charges against Zapata, pointing to 11 criminal violations.[11] Among these were robbery and destruction of haciendas; kidnapping; destruction and robbery of stores; and robbery of a factory, in the process of which a French worker was murdered and the administrator tortured. Readers of the *El Imparcial* article were probably most repulsed by the brutality represented in the tenth and eleventh charges. The tenth charge reports that Zapata forced a Spaniard to kiss the corpse of a soldier that had been decomposing in the middle of the road for ten days. The eleventh reports that Zapata laughed as he burned alive soldiers who had taken refuge in a rail car.

In the enumeration of charges, *El Imparcial* moves toward the development of Zapatista criminality as a character flaw manifested most clearly in the barbaric and excessive nature of Zapata's revenge.[12] Criminal anthropology at the turn of the last century, as developed by Cessare Lombroso, an Italian physician, used Darwinian concepts of the connection between human beings and other animals to develop a physical and psychological profile of criminal types (Lombroso-Ferrero 1911:72). Lombroso argued that certain types of individuals represented physically and mentally an earlier evolutionary stage of human development (Wilson and Herrnstein 1985:72–76). Lombroso identified physical features that supposedly resembled more primitive human types: long arms, sloping foreheads, and heads too small or large (ibid.:73). Humans who fit this physical type were born criminals.

For Mexican elites, Lombroso's work effectively marked the distance between normal society and the criminal world, underscoring the dimensions of terror posed by Zapata and the Zapatistas. Zapata's capabilities as a fighter became attributed to his animal-like cunning and his capacity to live and function in harsh environments. As relayed to the Mexican public by *El Tiempo,* the very successes of the Zapatista movement leaned upon the criminal character type of Zapata and his followers:

> The criminal type is a distinct individual that can be recognized among other reasons for his harsh physique which permits him to tolerate pain, hunger, sickness that would cause another man to succumb. Furthermore he is endowed with an agility that allows him to make a mockery out of any attack. Thus, the criminal is untiring, cures himself of all injuries without professional medical help, and is able to live deprived of any help or sustenance. . . . In order to triumph in the fight against criminality, the society should not emphasize the criminal act, as do existing laws, but rather the personal characteristics of delinquents—the things that make them so frightening. . . . In a word, it is not necessary to await another incident, just as one does not await another attack of insanity to bring a person to the mental institution, but rather having demonstrated the existence of a criminal character type the defense of society requires that prophylactic measures be taken to avoid continued criminal exploits.[13]

In essence, if Zapata fit the criminal type then hard proof that he was connected to crimes was unnecessary. Reporters, promising to return with "unbiased and objective" accounts, sought out Zapata in the hopes of exposing the deep personal and psychological profile that would make sense of his power to terrorize. They knew to look for evidence of paranoia, signs of character instability, and abnormalities in social attachment. To these qualities they added the unique cultural sensibilities of the Mexican elites.

Reporters returned with evidence of Zapata's criminality drawn from the physical environment in which he lived, the family and friends that surrounded him, and the culture and customs one abandoned to contact him. Out of these reports emerged a picture of Zapata as insane rather than as endowed with animal-like abilities. Thus, reporters play a pivotal role in trans-forming Zapata from an animal like-monster with unique abilities to survive into a human being whose behavior can be explained using the reigning dis-courses of scientific knowledge. Subjected to their gaze and analysis, Zapata became criminally insane, his power to terrorize brought to human dimen-sions while his ability to carry out a reasonable political project was denied.

On January 14, 1912, *La Prensa* carried pictures of one of their reporters with Emiliano Zapata and a signed letter from Zapata verifying that he had indeed met with and given the reporter an interview. The headlines quoted Zapata: "I would prefer to die rather than surrender; I will respect the poor but will be inflexible with the hacendados and especially with the Spaniards." The colorful description that follows traces the *La Prensa* reporter's frightful journey into Zapata's world, beginning with his search for the famous bandit:

> I headed for the house of Senora Espejo, the mother-in-law of Emiliano Zapata . . . she appeared at the door. I presented my credentials and ex-plained my purposes and she invited me in casting a look of suspicion over my leggings as I entered. Her mistrust heightened when I explained what I was looking for: information about the whereabouts of her son-in-law. She answered that she knew nothing and had nothing to do with him. . . . Her eyes betraying a tremendous hate which indicated to me that she had to know everything but she limited herself to telling me to go to the Mu-nicipal Mayor . . .

The mayor tries to find the reporter, Aldo Baroni, a guide; but few want to take the trip—they tell him that the area is full of Governor Figueroa's troops. Baroni learns that these frightful troops often attack honest, hard-working people, changing their uniforms to appear as Zapatistas while they rob and pillage. Finally, with an escort, he travels through the deserted state of Morelos until arriving at the hacienda Chinameca. The hacienda figures in Baroni's narrative as the last refuge of civilization:

> My arrival is greeted with a question that fills me with satisfaction, "and your stomach?" The question comes to me from the administrator of the hacienda before I have even dismounted my horse. "Breakfast or lunch?" "lunch" I answer with enthusiasm. In San Juan Chinameca life goes on peacefully, the monotony broken only by one or another false alarm.

Food—how much is eaten, how it is eaten, and how it is shared—becomes a symptom of mental and social health. The reader will recall, ac-cording to the account in *El Tiempo,* the criminal physical type can withstand

hunger as well as other hardships. To inquire as to the state of Baroni's stomach is to acknowledge that which Zapata will prove unable to recognize.

Before leaving, the reporter enjoys a friendly game of poker with the hacienda employees, a game that he will repeatedly look back to as his last pleasures of civilization. He leaves the hacienda, stopping at a ranch of five or six
"miserable shacks" in which no one seems able or willing to provide him
with information about the Zapatistas. He talks to the person who appears
to be the "head of the tribe," who is shocked to discover that newspapers are
still being printed, a misinformation that he believes must be due to Zapatista propaganda. Finally, in San Pueblo, Baroni's guide learns that the
Señores passed by in the morning. The two head off:

> I advance slowly, sadly. Behind me with the Hacienda Chinameca I have
> left behind civilization, whose ultimate expression of social life can be
> found in the Casino of the workers and their innocent poker game. Ahead,
> danger, bad food, thirst, the unknown.

Baroni's ruminations are broken:

> Who lives? This voice awakens me from my ruminations. I stand in my
> saddle and yell with all the force of my lungs, "Reporter."
> "Dismount and don't move" . . . This is the rude response that comes
> to me from high up in the mountain covered with luxurious vegetation.
> Following the voice, a man descends cutting a path through the thickets
> with a machete. . . . In a moment I find myself surrounded (in front, on
> both sides, behind) men with machetes, sticks, pistols. Approximately 20
> people dressed in white [the traditional peasant garb] ask who I am and
> what I have come for. . . . I explain the purpose of my journey.
> Some remain behind to guard the area; the rest surround and escort
> me. Silence. My guide leaves me without asking anything; without saying
> anything. Arriving at the top of the mountain in an area free of vegetation,
> there we encounter approximately 70 poorly armed Zapatistas who sur-
> round me, glaring at me with looks of mistrust and hate. I understand that
> at that moment one poorly chosen word, one sign of weakness, could cost
> me my life . . .

As opposed to the welcoming warmth offered by the administrators at
the hacienda and even the local mayor of Villa de Ayala, in Zapata's world
the reporter encounters silent people whose stares he must read. The silences
speak in a constantly shifting interpretive terrain, which Baroni deftly travels, invoking a series of frightening possibilities: silence as the resistance of

Zapata's mother-in-law who knows but will not speak; silence as custom, manifested by his escorts; silence as betrayal by the guide who abandons him without word; and the questionable silence that might signify stupidity or refusal as displayed by people he visits in the *ranchos* while searching for Zapata. Silence is particularly threatening to Baroni: people do not seem to respond to his credentials or his requests, nor do they understand anything about the existence of newspapers. It is as though Baroni has uncovered the limit of governmentality in his journey.

Baroni is escorted to Zapata by one of his generals:

> A little later I found myself seated at a rustic table in front of a cup of chocolate while Don Emiliano Zapata ate a plate of beans . . .

Descriptions of Zapata's eating habits surface frequently in newspaper articles probing the revolutionary leader's character. At issue is the kind of explanation offered for Zapata's meager diet. If he enjoys animal-like traits he may not need to eat, but if Zapata refuses food because he is consumed by a character flaw, such as paranoia, he might be merely insane. Baroni's explanation points to the latter. He writes:

> Zapata spoke little like most rural people, and I was not able to get much from him on this first meeting. I was able to note a bit of the trait that suggests his great mistrust. Upon finishing eating I offered him a cigarette, pulling one out for myself as well. Zapata held the cigarette in his hand for a moment staring at me with obstinate hate. He returned the cigarette to me taking mine from my hand and lighting it. At my expression of surprise he answered, "So that it blows up on you if you have put anything in it." I smoked the cigarette, that was only an innocent cigarette, and went to sleep at the order of the revolutionary . . .

Discussions of Zapata's great mistrust echo Baroni's description of Morelos peasant villagers who are always silent and mistrusting. Zapata's meager diet and his reluctance to share food become a symptom of his paranoia. Indeed, upon his assassination (accomplished through trickery) at the hands of government troops the newspaper *El Pueblo* reported:

> . . . Once in Cuautla, the body of Emiliano Zapata was laid out upon some crates to be examined by the public. . . . We should report as a curious fact that in the autopsy carried out before preparing the body, it became apparent that Emiliano Zapata had only ingested liquids, proving that the revolutionary leader was surviving only on *atole* (chocolate drink) for fear of being poisoned.[14]

Zapata's eating habits were contested in a number of newspaper articles. A reporter from *El Diario del Hogar,* a pro-Zapata newspaper, undertook a journey to Zapatista headquarters to interview the revolutionary leader and described in detail the luscious five-course meal he was served.[15] Zapata himself, in interviews that he granted, emphasized a normal concern with food and the toll deprivation had taken on his body:

> When I took up arms I did so for patriotic reasons and not in pursuit of personal interests. . . . In Cuernavaca . . . I enjoy all the fruits of my land and labor . . . I have ample land on which I cultivate watermelon and melons (fruits that I have to buy here at a high price. . . . They [hacendados] are repulsed by me because the people that they held as slaves have acquired their liberty by joining me in revolt. . . . Now, as I have said to you, I want to retire to a private life with the idea of curing myself and attending to my own interests . . . I am thinking of going to Tehuacan because I am suffering from frequent bouts of hepatic colic.[16]

Zapata's self-representations, as well as those manufactured by the few supporters he found among the mainstream press, emphasized an individual who could easily fit within the hegemonic elite social order, or at least with the way elites imagined themselves: He appreciated a stable full of fine horses, hard work, and the tranquility of family life.

More generous portrayals of Zapata were countered by reports that Zapata's personality was marked by horrendous mood swings reflected in inappropriate, uproarious laughter. Zapata, reportedly a man of few words, is prone to outbursts of diabolical passions. Baroni reports:

> . . . [Zapata queried] me as to what was being said about him in Mexico and interrogated me regarding troop movements. When I told him that very shortly he would be pursued by a tremendous column of federal troops resolutely intent upon putting an end to him and to the Zapatistas, he roared with laughter. Extending his finger toward Cuautla, the troops' headquarters, he yelled in a terrifying voice, "Put an end to me—but first you must put an end to the whole state of Morelos . . ."

Baroni describes yet another surprising outburst on the part of Emiliano Zapata during their second meeting. The meeting takes place as the Zapatistas begin to move into place for an attack on federal troops:

> . . . I and my guards headed along the road toward Los Hornos. During the journey my escorts boasted interminably. . . . With the setting of the

sun we arrived at a modest house of palms in Los Hornos. . . . We headed toward the patio. . . . Before my companions had any time to explain, a well-known voice bellowed from the crowd, "It's that shameless reporter" . . . Emiliano Zapata approached me. . . . receiving me with an affectionate embrace. . . . If I were to live a thousand years, I would never forget those two days of my life that I spent at Zapata's side—a kind of legendary life that one only encounters in adventure books: life of pirates of the land and forest, life that reminded me of childhood stories: Mandarin, The Tiger of Alicia, the Bandits of the Australian Mines . . .

In this liminal space where the norms of polite society are inverted and passions seem to run wild, Baroni rediscovers his childhood fantasies. He betrays a dash of romanticism for the criminal life that belies all the boundaries that he has erected between his own world and that of Zapata's. But, that Zapata's life is not childhood fantasy becomes clear as Baroni listens to Zapata and his men making war plans:

As the reports continued to arrive, Zapata began to dictate orders for the dispersion of his forces. His orders made me shiver with fright as I thought of the fate of the poor federal soldiers. Zapata led me to believe that he had at his disposition more than two thousand men. . . . I would have liked to have been able to take flight to warn a friend, sleeping soundly just a few kilometers from me without knowing that just an hour away the Atila of the South was preparing his fate.

Here Zapata proves capable of a curious merger of insane passion and rational planning. Zapata uses not animal-cunning to shape his attack but what, in Baroni's pen, becomes a coldly rational mind wedded to criminal insanity. Perhaps even more symptomatic of his insanity is his ability to use trickery and deceit.

Awakened from a shallow sleep, Baroni finds himself, as he had feared, part of Zapata's forces as they proceed for an attack. He complies with the orders delivered by the Zapatista forces without asking any questions, for, as he explains, "to speak in such difficult moments can be dangerous, and is always useless." He squeezes his pistol. But the frightful battle never really takes place:

Zapata ran when the enemy approached, ran shamefully into the mysterious night when confronted with the tireless push of the tiny, heroic band of federal soldiers. . . . The combat dispositions were all a farce worthy of Tartarin de Tarascon. . . . Zapata had no more than 200 men in Los Hornos.

Baroni's journey with Emiliano Zapata served to suture the space between the attacks of Zapatistas on trains, haciendas, and hotels and Zapata as an individual. If reporters could not prove that Zapata carried out these attacks they could at least demonstrate that his character was of such a frightening combination of cunning, unpredictability, and paranoia that only he could have masterminded such attacks. Ultimately, interviews both pro and con centered upon the question of Zapata's character.

Just as the terror of the attacks underscored the legitimacy of elite society, so the standards of Zapata's character were set in relationship to elite norms of sociability in dialogue with criminal anthropology. What did he eat and how often? How did he entertain his reporters, and what did his outbursts signify? In contrast to Baroni's report of his stay with Zapata, elite society was characterized by reporters as constructed around easy and trustworthy human relations marked by an appreciation for the good life invoked most often in the image of sharing food, drink, and cigarettes. Zapata's respectability, his appreciation of good food and fine horses—his traits that later became legendary—were, during the revolution, undermined with tales of his unpredictable character and his susceptibility to insane outbursts.

The focus on Zapata's character flaws as evidence of his criminality recalls other narrative treatments of criminality among the disenfranchised. For example, Slotkin's (1973) examination of narratives concerning African American crimes in New England between 1675 and 1800 speak to a preoccupation of Puritan New England with African American sexuality that became coupled with fears of organized revolt or rebellion. Ultimately, the linking of character flaws to fears of rebellion speaks to the symptoms associated with the emergence of *homo criminalis*. Here the marshalling of evidence of deep character flaws links individuals to class and racial categories to create the specter of a widespread eruption from below. In place of the examination of the individual criminal as a medical and legal phenomenon (see Foucault 1975), a politicized criminality highlights a pathological underworld social order that soon might come to power.

CONCLUSION

While the documentation and narration of criminality always takes place in a political context, the narration associated with establishing the criminality of acts that might be viewed as political or revolutionary is marked by a specific tone. The establishment of Zapatista criminality, through newspaper accounts of the revolution, drew on reigning ideas about the emergence of a criminal subject. Reporters pointed to Zapata's physique, the form of his laughter, his food consumption habits, and his moods to construct his criminality. In these areas, Zapata was shown to share traits with other common criminals. In another sense, Zapatista criminality differed from the scientific definition of criminality; the narration of Zapatista criminality was as much about defending elite culture as it was about examining a criminal subject. As a result, Zapata and the Zapatistas rarely appear as the center of their own acts, for those acts were only intelligible through the lens of the lives affected by them. Within the overall domain of crime, narratives of Zapata's criminality highlighted the important role of the state in protecting life in detail.

Foucault emphasizes governmentality as emerging in the context of relative peace, but here I have argued that a special kind of celebration of governmentality takes place in the reporting of revolutionary violence. At such times the reporting and documentation of allegedly criminal acts becomes secondary to the representation of the toll those acts take on governmentality. Indeed, narrative reports establish criminality through a dialogue between acts and the life in progress of those who stand for civility, the latter best understood as the manifestation of a form of self-governance. Furthermore, such reports reveal the limits of governmentality by representing the spaces within a sovereign nation where self-governance fails. Those who are not deemed worthy of self-governance, such as the peasants visited by Baroni, lack daily routines that might be disrupted. Peasants are represented as entirely outside the rubric of self-governance, and the loss of their lives is rarely lamented.

I suspect that in situations where acts could be constituted as either criminal or political, the establishment of criminality depends upon constructing narratives in which criminality is refracted through the lens of the disruption of everyday routines. These narratives must be read against the grain of criminality for they provide a window into those routines of life that are treated with special reverence. By attending to the systematic celebration of some routines and the equally systematic silence regarding others, the narratives speak to the structures of power through which criminal identities are constituted.

ACKNOWLEDGEMENTS

This research was supported by a Professional Development Fund Grant from Earlham College. I am especially grateful to the librarians in the newspaper archives at the University of Texas, Austin, who gave me full access to the collection and to Philip Parnell, who provided insightful feedback on this paper.

NOTES

1. This article draws on research still in progress in which I have been combing articles written about Zapata and Zapatismo between 1910 and 1919 in newspapers from central Mexico.

2. For a discussion of "local" memories of Zapata see Martin (1993) and Brunk (1995).
3. All translations of newspaper articles are my own.
4. *El Imparcial,* June 19, 1911; *La Prensa,* March 31, 1912.
5. *El Diario del Hogar,* October 10, 1911, criticizes Madero for appointing Zapata's enemy, Figueroa, governor of the state and argues that Zapatista forces are growing because *hacendados* have been allowed to trample upon the rights of peasants.
6. The fear of Zapata assuming a position of power in Madero's government was particularly acute in 1911 when planters in the state of Morelos sought to direct the revolution in their own interests. They complained that Madero was not being harsh enough on Zapata and sought federal intervention to control Zapatista troops. Indeed, Madero appears to have resisted the planters' most outrageous depictions of Zapata while never truly supporting Zapata's demands for land reform (*El Imparcial,* June 19, 1911). See also the discussion by Alan Knight (1986: 260–264).
7. *The Mexican Herald,* August 18, 1911.
8. *El Imparcial,* August 27, 1911.
9. *The Mexican Herald,* August 29, 1911.
10. *El Imparcial,* August 17, 1911.
11. Ibid.
12. *El Tiempo,* December 9, 1911.
13. *El Tiempo,* December 19, 1911.
14. *El Pueblo,* April 12, 1919.
15. *El Diario del Hogar,* May 19, 1911.
16. *El Heraldo de Mexico,* June 21, 1911.

BIBLIOGRAPHY

Anderson, Benedict
1983 Imagined Communities: Reflections on the Origin and Spread of Nationalism. London: Verso.
Bazant, Jan
1977 A Concise History of Mexico from Hidalgo to Cardenas, 1805–1940. Cambridge: Cambridge University Press.
Brunk, Samuel
1995 Emiliano Zapata: Revolution and Betrayal in Mexico. Albuquerque: University of New Mexico Press.
1996 "The Sad Situation of Civilians and Soldiers": The Banditry of Zapatismo in the Mexican Revolution. American Historical Review 101:331–353.

Comte, Auguste
2001   The Positivist Philosophy of Auguste Comte. Harriet Martineau, trans. Bristol, England: Thoemmes.
Foucault, Michel
1975   I Pierre Rivire, having slaughtered my mother, my sister, and my brother: A Case of Parricide in the 19th Century. Lincoln and London: University of Nebraska Press.
1990   The History of Sexuality: Volume 1: An Introduction. New York: Vintage Books.
1991   Governmentality. *In* The Foucault Effect: Studies in Governmentality. Graham Burchell, Colin Gordon, and Peter Miller, eds. Pp. 87–104. Chicago: University of Chicago Press.
Gordon, Colin
1991   Governmental Rationality: An Introduction. *In* The Foucault Effect: Studies in Governmentality. Graham Burchell, Colin Gordon, and Peter Miller, eds. Pp. 1–51. Chicago: University of Chicago Press.
Knight, Alan
1986   The Mexican Revolution: Volume 1: Porfirians, Liberals and Peasants. Lincoln: University of Nebraska Press.
Lombroso-Ferrero, Gina
1972 (1911)   Lombroso's Criminal Man. Montclair, N.J.: Patterson Smith.
Martin, JoAnn
1993   Contesting Authenticity: Battles over the Representation of History in Morelos, Mexico. Ethnohistory 40(3):438–466.
Miller, Robert Ryal
1985   Mexico: A History. Norman: University of Oklahoma Press.
O'Malley, Ilene
1986   The Myth of the Revolution: Hero Cults and the Institutionalization of the Mexican State, 1920–1940. Westport, CT: Greenwood Press.
Slotkin, Richard
1973   Narratives of Negro Crime in New England, 1675–1800. American Quarterly 25:3–31.
Wilson, James Q. and Richard J. Herrnstein
1985   Crimes and Human Nature. New York: Simon and Schuster.
Wolf, Eric
1969   Peasant Wars of the Twentieth Century. New York: Harper & Row.
Womack, John
1968   Zapata and the Mexican Revolution. New York: Vintage Books.
Wilson, James Q. and Richard J. Herrnstein
1985   Crimes and Human Nature. New York: Simon and Schuster.

CHAPTER 8

CRIMINALIZING COLONIALISM
Democracy Meets Law in Manila

Philip C. Parnell

THIS CHAPTER IS ABOUT THE PROCESS OF CRIMINALIZATION as it arose in the struggle for control over a region of urban land among Metropolitan Manila intergroup networks. Disputes over land intensified in the resurgence of democracy that followed the Philippine People Power Revolution and the disintegration of President Ferdinand Marcos's dictatorship. Network participants used the criminal identity, which had not been salient in intergroup disputes, as they created narratives of nationalism that legitimized both their rights to land and their ways of occupying and controlling vast tracts of urban real estate. As components of these networks, officials in the government and the legal system as well as religious leaders actively participated in using them to develop their versions of the Philippine state and nation. Manileños put their views of the state and nation into action through the different ways they composed intergroup networks, structured participation in them, and developed network strategies for gaining permanent residence on urban land. The process of criminalization emerged for the first time in disputes across these and other intergroup networks as officials of the church and state became more involved in their disputes. Their participation, along with the end of dictatorship and the ascension of people power, seemed to create clearer demarcations among networks. Although these distinctions were based on differences in network alliances, they elided similarities across networks and ways their members were associated. Some networks, represented here by those called land syndicates, most often pursued land ownership through the legal system and corporate bureaucracies while constructing the nation and the state out of colonial law, arguing that the contemporary Philippine state was unstable and unpredictable. Other networks, represented in this discussion by Sama Sama, a people's organization, pursued land ownership through democratic processes and links to religious and governmental organizations while talking of a nation that prioritized the interests of Filipinos over Western notions of economic development.

I draw on my ethnographic research in Manila to explore why some intergroup networks, as enactments of the Philippine state and nation, were characterized and treated as criminal while other similar networks were not. A possible answer that I consider below lies in the nature of Philippine sociality and the very mechanisms through which Filipinos can construct their own versions of the state and nation out of the vicissitudes of survival in a changeable world; out of their interpretations of the Philippine colonial past; out of the more recent experience of dictatorship and martial law under former President Ferdinand Marcos; and out of the 1986 People Power Revolution that signaled the national resurgence of Philippine democracy. In brief, the category of criminal became salient following the People Power Revolution when land syndicates were characterized as representing interests that were not truly Filipino—as a barrier to a Philippine nation. Although I would not attempt to and could not capture how being Filipino is constructed in all situations, from the ethnographic viewpoint I gained while living in Manila a preferred way of relating to society entails a structural disencumbering through which individuals may maintain multiple and even contradictory associations and affiliations; this is a form of identity formation that prioritizes the processes of associating over the substance of affiliation. Intergroup networks encoded this process through both rhetorics of nationalism and processes of participation. With preserving this agency in the context of change in mind, urban poor Filipinos evaluated intergroup networks as differing mechanisms of social participation and survival as well as differing ways of enacting the Philippine state.

The People Power Revolution, which led to the flight of former dictator Ferdinand Marcos to Hawai'i and the election of President Corazon Aquino, who championed democracy, shifted power and perception across intergroup networks in two ways important to this discussion of crime. Those networks that had arisen from and championed grassroots democratic action, such as Sama Sama, saw their practices and civil discourses formalized as the official ideology of the Philippine national government. In this, they were empowered and legitimized by a process that Pierre Bourdieu (1987), discussing the "Force of Law," calls homologenation (the force of homology). State law achieves the "homologenation effect. . . . by ordaining the patterns that [already] govern behavior in practice . . . through the objectivity of a written rule or of an explicitly expressed regulation" (ibid.: 849). As this effect empowers law, giving it the appearance of truth among those it mimics, power also travels in the opposite direction—the notion of being right blossoms or is nourished among those whose practices become expressed as law. Urban poor members of Sama Sama, whose mechanisms of survival were reflected in official state policy following the People Power Revolution, through the force of homology began to see themselves somewhat as the only truly Filipino (democratic) network, rather than as, previously, one option among many networks, including land syndicates, that could provide residential stability to the poor. Their vertical ties to the state (and the church) assumed greater significance amidst their many horizontal alliances with other urban poor networks.

Contemporaneous with democratization and the creation of popular national homologies of grassroots democracy, another process—iconization—strengthened ways in which intergroup networks perceived each other as opposed rather than interconnected (as they were through interrelated members and some shared ideologies). I draw on the linguistic concept of iconization developed by Irvine and Gal (2000) to designate the process through which perception of a person or group's behavior and the ways they are characterized by others become the same. As Sama Sama enjoyed the fruits of homology, land syndicates were weakened as, in the eyes of others, they came to symbolize past challenges to Philippine nationalisms. Although land syndicates were structured as intergroup networks and did not limit members' affiliations to syndicate leaders and groups, they, unlike networks with grassroots democracies at their bases, were structured as bureaucracies. They distributed land to individuals from a single owner legitimized in colonial law; attempted to adjudicate their land claims through the official legal system; sought legitimacy, in part, through a rhetoric of alignment with the United States; championed international business associations; and characterized themselves as both private corporations and hierarchical governmental organizations. Most importantly, land syndicates sold themselves to the public through their associations with official Philippine law that governed the distribution of land. In the context of democratization, these networks were increasingly perceived by grassroots democrats less as multifaceted conglomerates that provided some residential stability to the poor, and more as authoritarian bureaucratic structures through which their internationally linked leaders attempted to adjudicate land disputes for their own enrichment. Seen in this way, land syndicates challenged both nationalized and local notions of democratic rule that placed authority within the state in the hands of Filipinos; on the other hand, the homologized Sama Sama was a truly Filipino path to gaining and maintaining a multiplicity of relations necessary to survival in a changing world.

In this context of homologies and icons that emerged in the process of Philippine democratization, the process of criminalization took hold, somewhat, in relations among Philippine intergroup networks and without the invocation of official law. Land syndicates were accused of criminality by urban poor members of Sama Sama, by governmental agencies linked to Sama Sama, and within the Philippine Congress. Syndicates resisted this characterization by asserting their rights to sell land and by expanding their links to other networks and a wider range of groups. Some Filipinos continued to view syndicate leaders as national heroes and respected patrons.

The criminalization of syndicates was more than a rhetorical strategy in disputes over the best way for the urban poor to gain both the ability and right to live in Manila in the context of others wanting the land they occupied. In the example I offer below, the criminalizing process became actionable—rather than a matter of talk, lists, and accusations—against syndicate-related groups that did not practice multiple affiliations. It is important to note here that, for the urban poor, democratization did not eliminate challenges to survival. The power of criminalization was mitigated by a syndicate's demonstrated ability to provide and secure land tenure for Philippine households.

A CONTESTED URBAN REGION

I draw these notions of popularly constructed and actionable criminality from ethnographic research in Payatas Estate, a more than 2,000-acre region of Quezon City (the capital of the Philippines), which is part of Metropolitan Manila. In addition to hosting the National Government Center (a region of Payatas claimed by the Philippine government where the Philippine Congress is located, also called the NGC) and around 150,000 squatters (in 1987 and 1988, the time of the research I consider here; the squatter population has grown over the years), Payatas is the location of several headquarters offices for federal governmental agencies and a few small private subdivisions. During my initial research in Payatas, I lived for 11 months in a one-room shanty located in the National Government Center. I have often thought of Payatas Estate as a living memorial to some of the ways Filipinos, over more than a century, have attempted to gain and distribute land among groups and individuals, mostly the urban poor (who comprise most of the population of Metropolitan Manila), who were searching for residential stability as well as economic capital while resisting relocation by federal and municipal governments and private real estate corporations.

In addition to localized homeowners' associations, numerous larger city-wide and national organizations emanated from Payatas Estate. I focus here on three of these intergroup networks: land syndicates, Sama Sama, and Samako. The notion of criminality arose among these organizations as they were expanding their memberships and seeking to encompass a wider range of groups in their networks. Criminality was linked only to Payatas land syndicates and not, to my knowledge, to other syndicate forms. Although there was widely distributed criticism of other groups called syndicates that emanated from governmental bureaucracies and that appeared to interpenetrate official and unofficial processes of land distribution and titling, they, to my knowledge, were not characterized as criminal. The Philippine Solicitor General, in an interview I conducted, bemoaned syndicate activity within governmental bureaucracies without suggesting they were criminal or susceptible to criminal prosecution.

Land syndicates, Sama Sama, and other intergroup networks represented in Payatas were structurally similar. At their bases were those who occupied the land they sought to control, mostly squatters and the urban poor. These participants were divided into numerous other groups, such as homeowners' associations and regional chapters (in the case of Sama Sama). Payatas residents were characteristically aligned with more than one intergroup network and their group memberships were not confined to Payatas. Some individuals interconnected networks through membership in more than one. These populist bases were then connected to what I will call a network's primary institutional forms. For syndicates, those forms, consistently hierarchical and bureaucratic, were the private corporation, the official legal system, and the military. All syndicate leaders had claimed to be heirs to Spanish land titles that cover all of Payatas (as well as much of Metropolitan Manila), or the heirs' appointed administrators of the land. The syndicate leaders most active in Payatas claimed to be the true government of the Philippines.

Sama Sama, whose active members were mostly women, maintained primary institutional connections to the Catholic church and the executive branch of the federal government, represented mostly by those appointed to leadership positions within housing bureaucracies by President Aquino. Sama Sama's alignment with the church connected it to numerous nongovernmental organizations (NGOs), politicians, and academics. Sama Sama, a democratically governed network linked to numerous other urban poor organizations, championed grassroots development, democracy, and local autonomy.

The founding members of Samako, a mostly female organization that championed formation of land-based economic collectives, had been among the original members of Sama Sama; but, as the group took form, these members had aligned with a Marcos-controlled housing bureaucracy to gain influence over land distribution. Winds of political change within the Catholic church that preceded the People Power revolution loosened Samako's ties to church sponsorship. Searching across networks for supporters among the urban poor, Samako refused to bend politically, instead increasing its distance from its offshoot in Payatas—Sama Sama—and many of the NGOs, political networks, and grassroots supporters of the urban poor that mushroomed following the 1986 resurgence of Philippine democracy. Isolated, Samako eventually sought land through one of the larger syndicates and became the target of anti-criminal action.

I originally moved into Payatas to study disputes across groups that identified themselves as different, assuming I would find relatively exclusive associations based on members' identifications within geographic (or, if they chose, cultural) regions of the Philippines. However, such regional fracturings and oppositions that analysts have attributed to the Philippine elite (McCoy and De Jesus 1982) were not characteristic of the urban poor in Payatas. I realized this as I spent my initial months in Payatas traveling across the settlement and Metropolitan Manila, from meeting to meeting, house to house, and office building to office building with members of Sama Sama. My approach to ethnography was phenomenological (see Jackson 1996): Because I felt I could assume little about the lives and thoughts of Payatas' squatters, I relinquished control over my time and quietly attached myself to the daily activities of group members. To gain a broader view of settlement social life, politics, and disputes, I frequently joined lengthy evening discussions among small groups of my male neighbors meeting in their yards.

Traveling through the day with members of Sama Sama, most of the time I did not know my destinations. At times, awakening early, I merely began walking through the settlement on my own until I encountered Sama Sama members and joined their activities. My first impressions of Manila squatter settlements would not have allowed this ethnographic approach: From the outside, they appeared unorganized, deteriorating, densely packed, incredibly poor, and intimidating. In spite of my cultural training, these qualities raised my Western concerns about danger and crime. Once I became a resident of the sprawling Payatas squatter settlements, however, I discovered that their unpaved, often muddy streets were safe to the point that both day and night I could walk the unlit streets without fear. Although there were, of course, interpersonal relationships that would have been la-

8.1. The author's one-room shanty in Manila and other squatter homes in the National Government Center were sites of contestation that energized some of the Philippines' most powerful peoples' organizations.

beled criminal in the United States, such as robbery and forms of domestic disputes, crime was not a salient category among the urban poor of Payatas.

Some of my interactions through Sama Sama involved their ongoing recruitment of members by linking to other associations; membership expansion was necessary to Sama Sama's choice to create a democratic base in Payatas. I could not, however, sufficiently understand these interactions without gaining the other groups' perspectives. Expanding my research to include groups throughout the Payatas region, I consciously entered the turf of the land syndicates. I use the term consciously here because although I had previously interacted with syndicate organizations, Sama Sama, practicing a studied silence, had never identified these organizations as components of syndicate networks. For example, after attending a meeting in one of Manila's business districts among Sama Sama leaders, representatives of its church alliance, and representatives of the Aquino government, I traveled with Sama Sama to the nearby offices of a private corporation. There we sat around a table with the corporation's leaders, who included a former appellate court judge and an academic known for his associations with urban poor causes. They offered an alliance with Sama Sama if they would incorporate in their planned housing project use of prefabricated concrete housing the corporation was importing from Korea. The corporate leaders seemed to suggest that this alliance was necessary for the project to move forward although, as anyone who lived in Manila knew, the urban poor could easily construct their own concrete houses, and, in doing so, create jobs for unemployed Filipinos, a goal of Sama Sama's do-it-yourself housing project. After the discussion, members of Sama Sama asked me what I thought of the exchanges, without revealing their knowledge that the group was, as I discovered later, closely associated with one of the larger syndicates operating in Payatas. The judge was known for his finding in a syndicate initiated case that a syndicate leader was indeed the heir to the Spanish title to all of Payatas' land.

To me, the syndicates, as associations comprised of homeowners' linked to people of influence, had been indistinguishable from other groups. Members of Sama Sama related to other urban poor who were in syndicates as members of the groups that comprised the networks, such as homeowners' associations, rather than as members of syndicates. Syndicate leaders were also widely viewed as champions of the powerless. Some organizations that had been identified by the state as syndicates, as well as some urban syndicate leaders, had been associated with efforts to achieve rural land reform. Many landless throughout the Philippines sought such reform, viewing the state's past alliances with foreign corporations as robbing Filipinos of their own land. Some land syndicates also defended the interests of penny capitalists who operated small stalls in areas frequented by customers of larger stores and malls, especially when their presence was challenged by the larger retailers. The populist identity of syndicates was more than rhetoric, and the networks provided and promised entrance into more secure economic roles for many Filipinos. Many members of Sama Sama had originally acquired land in Payatas through the syndicates, mostly during the Marcos presidency. It is my understanding that they did not view the gaining of land to occupy through syndicate ties as illegal. No members of Sama Sama held legal titles to the land they occupied in Payatas. But they constructed their rights to land in political and economic contexts rather than through state law. Syndicates were not viewed by the urban poor as criminal for facilitating the rights of Filipinos to use unoccupied or undeveloped Philippine land to pursue livelihoods and stability for their families. Also, some members of Sama Sama families made their livings by working within organizations that could be or were labeled syndicates.

The term syndicate, then, was mostly applied to the larger network and its leader rather than to its members or component groups. People spoke of their activities through the names of these groups, the names of leaders, and the names of members rather than through the term *syndicate*. But the term land syndicate achieved greater currency in Manila in 1987, and members of Sama Sama began to openly discuss some of their neighbors as syndicate groups. Disputes between Sama Sama and the syndicates were escalating as Sama Sama sought alignments with components of syndicate networks. The many ways in which Sama Sama and the syndicates were interrelated began to become less visible.

My research on syndicates by happenstance coincided with Sama Sama's shifting of its major organizational activities from the National Government Center portion of Payatas into regions where syndicate control of land was strongest. Ironically, just after my planning for research on syndicates had begun, the (Philippine) Presidential Commission on the Urban Poor (PCUP) published a list of syndicates and identified them as criminal organizations. Hearings on syndicates were held in Congress, where they were referred to as criminal. For the first time, the notion of a criminal had appeared in my research. As noted above, as intergroup networks, syndicates employed a common Philippine structural form, perhaps the dominant form within the decentered Philippine state. Structurally, they were very similar to Sama Sama. Syndicate leaders spoke of their networking as the same as the practices of Philippine state bureaucrats and politicians. But one of their most distinctive features among intergroup networks was how they presented themselves as Filipino—their nationalisms.

In the following, I explore some Filipino views on various kinds of organizations called syndicates then introduce syndicate leaders through their own words. I then further examine the contexts of their criminalization. Finally, that process culminates in governmental use of military violence against Samako as it found patronage within land syndicates.

"I AM BETTER BECAUSE I HAVE A DOCUMENT"

While following intergroup disputes emanating from Payatas, I occasionally asked people knowledgeable in Philippine approaches to land distribution and ownership to tell me what they meant when they spoke of syndicates. The replies were varied. Through discussions with commissioners at the PCUP and as suggested in its list of syndicates, most groups considered syndicates had challenged governmental ownership of land and/or subdivision developers that claimed to hold legal titles. Problematic to this characterization is that many other intergroup networks not labeled syndicates had also mounted such challenges. Most syndicates on the PCUP list were associated in some way with court cases involving Spanish land grants known as Título de Propiedad Número 4136 and Royal Decree 1–4 Protocol; both were issued around 1894 and had been recognized by the United States when it assumed authority over the Philippines in 1898. In a civil case initiated in 1971, and still in process in 1994 when I was last in Payatas Estate, several Payatas leaders claimed to be the owner of Payatas Estate as the legal heirs of one or the other title or through several transfers of deeds and authorities from the heirs. Both titles were claimed to cover all of Metropolitan Manila and had long been used by many of their self-proclaimed heirs to distribute land. A Congressional committee investigating syndicates included the same groups as the PCUP, also designating some syndicates (but not all groups popularly called syndicates) criminal organizations.

The director of the federal Housing and Land Use Regulatory Board, responding to my question about the nature of syndicates, replied:

> I was informed by the Solicitor [General's] office that [title] 4136 is null and void. I was informed that that was decided by the Supreme Court. Pasero [leader of one of the larger syndicates active in Payatas] does not sell the land but encourages people to occupy vacant areas in his claim and tells them "We will donate the land to you but we need to survey," etc. So he charges survey fees, administrative costs, and membership fees. I understand these people have genuine documents. I don't know how they were able to get forms from the Register of Deeds. They are legitimate forms. I don't know if they are printing them themselves. It works if they [the forms] look prestigious. A former Court of Appeals justice is now the lawyer for Pasero. The only way to tell if the title is real is by tracing it back. They are syndicates. They have people in the Registry of Deeds and the Bureau of Lands.
>
> The case [over title 4136] is now in court [in 1988]. The Estate, several years ago, was subdivided by the government and sold. After titles were issued, some buyers transferred rights and got transfer certificates. Third generation titles are now in the hands of buyers. Then someone filed a reconstitution alleging his titles over this land were lost. The certification presented to the judge was that the titles were not really issued, and the judge reconstituted them. On the basis of the certificate, the judge said there were no titles covering the property, and the person who issued the [allegedly false] certificate denies it.
>
> I was formerly with the Bureau of Lands [from 1953 to 1986] and the National Bureau of Investigation and saw all kinds of titles. They age documents because in some parish churches there is the same old paper that was used for Spanish titles [*papel de barba*], and there are people who can write in the old style and pass the paper over gas lamps for soot.

Concerning fake titles, one of the most successful syndicate leaders accused of being a criminal stated ironically, "But that is the case of the Philippines. People say never mind that the document is fake, I have the document. Do you have the document? I am better because I have a document, although fake, but I have a document."

The leader of a Manila-wide confederation of squatter organizations who questioned the legality of many titles issued by the Bureau of Lands, especially those issued to real estate developers, explained to me:

> Spurious title holders can get an order from Malacañang and use [federal] soldiers, pay them 300 to 500 pesos a day, and the [federal] soldiers will harass us. So we are ready for armed struggle if we have to. Might makes right, so if those with connections get right we may use might. These are not only judicial and legal struggles but also parliamentary and sometimes armed struggles if necessary against spurious title holders.

A leader of a squatter organization allied with the federal Bureau of Forestry Development, replied to my question about syndicates:

> There were members of the committee [a committee on land fraud formed by President Marcos] in the Bureau of Lands and the Land Registration Commission who had reservations because it is in their agencies where syndicates issue spurious titles. The syndicate is a group of employees who issue false titles to get money. They splice microfilms. Not all employees of the Bureau of Lands are part of the syndicate. The head of the Bureau of Lands is probably the head of the syndicate.

A congressmen who headed a committee that investigated the problem of syndicates answered my question stating:

> People who sell titles have inside contacts that steal security paper and produce titles. There have been many public land areas subjected to legal titling. People have built homes in subdivisions and discover the developer has false titles. I don't yet have final answers to this. Unless lands have been set aside for alienation they are government properties. The Payatas area is government property. In those areas where [syndicates] are not selling titles they are selling rights or access. They are not selling land but the right to be on the land. When claims to the land are settled, then they have the prior right of occupancy. Meanwhile, some of the titles are used as collateral in banks, and the bank will discover there are overlapping claims. . . .
> Somewhere something went wrong with the Marcos government. There was a loss of accountability. We have the baggage of dictatorship.

People exercised tremendous power simply because they claimed they were connected. Criminal things like this could have been easily squelched, but there was so much uncertainty. . . . But activities like fake titles remained underground. . . . Fake title syndicates are still there. They are not actually in governmental agencies.

I interviewed all of the syndicate leaders operating in Payatas Estate; all openly administered and distributed land, as did leaders of some of their constituent groups. All claimed that the government's and private claims to the land lacked legal merit. I first interviewed a syndicate leader while attending a birthday party thrown for him in Payatas by a veterans of World War II association that was linked to groups of homeowners (members were mostly children of veterans and their relatives who claimed the government had given Payatas land to the veterans after the war as a reward for their service). Among the guests were attorneys working for Quezon City, federal soldiers in uniform, and the regional chief of police. All syndicate leaders I interviewed claimed to have held connections to U.S. political leaders and the CIA. One had run for Congress on a platform that championed the Philippines as the 51st U.S. state. Most seek to legitimize their identities by citing periods of past persecution by the Philippine government resulting from their abilities to challenge its sovereignty on Manila's island of Luzon.

"WHAT I AM DOING IS THE JOB OF GOVERNMENT"

A portion of Payatas Estate borders Manila's largest reservoir and reaches down to the Marikina River. Near the river are two of Metropolitan Manila's largest garbage dumps. Don Emiliano Pasero lived in this expansive region of Payatas, where urban poor residents formed homeowners' associations that were part of Pasero's umbrella corporation called McManagement. Pasero described himself as the "private ancestral land administrator/owner" of Payatas Estate and other regions of the Philippines; some called him the Godfather of the urban poor. Pasero claimed to employ regularly 27 lawyers and to manage 250 corporations "all over the Philippines." He said he had registered corporations for mines and fisheries in addition to his other land and housing-related undertakings. Pasero explained, "Almost one million people (corporations and individuals) have received land from me. There are almost five million of my beneficiaries throughout the Philippines who will rally at Congress."

Don Pasero argued that many of those who held official legal titles to Payatas tracts acquired them illegally and that all titles issued to Payatas land since 1894 were illegal. Members of Pasero's homeowners' associations went through all of the steps of titling their land just short of receiving the (officially legal) Torrens title. At each step a governmental official was paid by the title seeker. In this process, governmental documents served as titles, although they were not officially legal titles. From Pasero's viewpoint, the documents were public testimony to their owners' willingness to align with the government and act within the official law, although the purchaser of the documents was not able to act completely within official law by also purchasing the Torrens title.

When I first talked with Don Pasero in his ranch-style concrete home, he offered his view of Philippine law:

> We don't have Philippine law now, we still have American law. So there is no law. The law of 20 years of (former President) Marcos is cosmetic. We can consider it cosmetic because (President) Aquino has removed it. We can't now determine if the law of Aquino is cosmetic because we don't know if it will be adopted by the people.
>
> I tried to handle some of the problems of land reform, but the law is crooked, so how can I help when those helping are not recognized because of the cronies, some of whom hired guns to cause trouble. The persons around Marcos disturbed him much. That's why my work of allotting land, in any Philippine province where I can allot land is being disturbed. I can allot the land alone, I can do it by myself. I have my 250 corporations; and if I think the people of the corporation are good, I will give them a big area for them to distribute to the poor at minimal cost or fee. . . . I want the poor to have land so they can have a farm to plant food.
>
> The hard thing for us now is that the government wants to bring all of the land I have allotted under operation of the law, which means I have to use courts to confirm or quiet a title. All of our people are poor. I have no more money to support them because it turned to land. Some billionaires force us off the land, and because we have no money we cannot bring it under the law.
>
> If the government always disposes the land without our consent, when another President enters all of the laws will be erased again and the land will come back to me. I am now amenable. The government cannot confiscate property without compensation and due process, that is my right under Article 29 of the Civil Code and Article 449 of the new Civil Code.

In our conversations, Pasero revealed he was well-schooled in changes that had occurred in land law since the end of Spanish colonialism, when the Philippines became a colony of the United States, through the martial law of Marcos and then the rewriting of the Philippine Constitution during Aquino's presidency. In his view, during each of these periods, official laws regulating the use of land were ignored. In the view of Pasero and many of the urban poor in Payatas Estate, who, through residents' associations, have studied the history of the titling of the land where they live, many of today's legal titles were derived from titles acquired or issued illegally.

Pasero's moral stance is that Philippine land belongs to Filipinos, and that should be the starting point for establishing the legality of land distribution and untangling the legacy of past colonialisms. He states, "The government must be strong so it can stand alone. If we use the land, we can pay back loans to other countries—if we plant the land. Some Filipinos and foreigners have taken lands by means of *haciendas*. If we plant, foreign visitors can see many fruits and take them home. We can recover our bad names."

There is irony but mostly humor in Pasero's statement that "the government wants to bring all of the land I have allotted under operation of the law, which means I have to use courts to confirm or quiet a title." For Pasero, official modern law is not yet legal; it is merely the latest of a series of legal systems that the people have not adopted. He can therefore legitimately use past law to negate that of the present: All are equally illegitimate.

Several other figures who claimed through the same court case that they were heirs to Don Mariano San Pedro's land headed Payatas land syndicates. Although title 4136 had never been ruled valid by an official court, a civil court, in title proceedings, appointed several of the heir-claimants or their representatives as administrators of the land and ordered them to dispose of its undeveloped portions. In disposing of the land in Payatas, the self-proclaimed heirs, whether appointed administrator or not, settled the poor on the land to help secure their control of it. Among the poor, police and military personnel were known to be effective defenders of homeowners' associations faced with demolition attempts by the government or private developers. Such attempts often pitted Philippine soldiers against each other in urban land battles. Syndicate leaders also awarded land to legal officials associated with municipal and federal legal agencies and courts.

Don Pasero draws a strong distinction between his challenge to official law and his support of the Philippine government. Through his methods, that is, working outside an official legal system that restricts the distribution of Philippine resources, Pasero claimed that he could unite the official government with the people:

> . . . what I am doing is not for myself but is the job of government; but, the government doesn't compensate me. I use my own self-liquidating process. I do the job of government because I am campaigning for land registration to educate people about the law of the land. Many don't know

the law, that's why the NPA (New People's Army), MNLF (Moro National Liberation Front), Yellow Army and military are used by private persons. Senators and congressmen are afraid to go to those areas. I tell the people to yield to the government and pay their taxes. Under me, they will not fight our government. We should pay taxes and help our government. Under others, they will not pay taxes, and they will fight. My security is to pacify the matter and bring it into the court. I go out into the countryside almost daily. Treasurer's offices in the provinces won't let them (squatters) pay for land because they don't have papers. I give them the papers so they can pay taxes. Sometimes government employees give them the land. I go to the people, talk with them, make the documents, check the documents with lawyers, and then file them in court. Treasurers and government employees collect money without papers in their possession. Treasurers can't go into the areas because they are afraid. Government officials are ignorant. They have education, but they don't know about the economic revolution of the government. All government money must be collected because the government has no industry.

"ALL GOVERNMENTS ARE SYNDICATES"

The syndicate leader Jorge Monarco set up office in Payatas Estate near the Philippine Congress while I was living there in 1988. Monarco claimed to be the rightful heir of Don Mariano San Pedro y Esteban and to be in possession of the original title 4136. Monarco explained that when he was 16 the Philippine government killed his father and tried to kill him because of their relationship to the title. He said that President Marcos, in 1972, burned down the town of Magsaysay, in the province of Bulacan near Manila, while in search of the title. Monarco told me that he had moved from place to place since he was 16 out of fear of the government:

> I can feel it whenever there would be some people who would like to pick me up. In our country now it is already ordinary to kill. They treat people as pigs. Just for a small amount they will kill. My house is already rotten but they will not still let me sleep soundly. Even though their houses are big, they cannot sleep for thinking of how to kill me.

Unlike Pasero, Monarco lived outside of Metropolitan Manila in a small two-room thatched hut with wooden floors that sat above the ground on stilts. A man of apparently little means who felt persecuted by the government, Monarco, through the intricacies of court proceedings over title 4136, had become the partner of the most recent court-appointed administrator of land

within the Don Mariano San Pedro Estate. In 1988, Monarco was openly selling occupied prized lots near the Philippine Congress building, some apparently within the government-owned National Government Center.

As I interviewed Monarco at his home, he claimed that he had various
land-sharing deals with governmental officials, including several Philippine
presidents:

> In 1980 I gave all of my documents to Marcos. The agreement with Mar-
> cos was one-half would go to the government and the people while the
> other half would be divided between me and Marcos. The land was issued
> during the first term of the Marcos presidency. Magsaysay (President of
> the Philippines from 1953 to 1957) died not by accident but because of
> an agreement he had made over the land, that Magsaysay had made with
> me. I tried also to make an agreement with Garcia (President of the Philip-
> pines from 1957 to 1961); but, when I heard that Garcia was ordering the
> killing of the heirs of Don Mariano, I didn't seek an agreement. Garcia was
> land-grabbing.

Like Monarco, Don Pasero created within his life history links to powerful figures and organizations: In one of our conversations he stated, "In the
newspapers and Congress they said I was supported by the CIA and Mafia
because I applied for a position with the CIA. Before, in (President) Garcia's
time, I was with Mao Tse Tung. At the time I was working as a janitor at
Malacanang (the Philippine presidential palace), and Garcia sent me to
China to learn about China. I was even in Russia. When I was suspected, I
went to the U.S., then back to the Philippines."

Monarco claimed to seek genuine land reform. In 1987, he wrote a letter to then President Aquino offering to assist in the development of urban
poor communities located on the land covered by 4136; he also requested
that the military assist his efforts. Monarco explained to me, "My objective
is to distribute the land to the people so they will not be called squatters. My
objective is that the Philippines is for the Filipino." Monarco claimed that
he was no different from those in power. He argued the Mayor of Quezon
City was "the biggest land-grabber of them all." Of the official Philippine
government Monarco told me:

> All governments are syndicates. They increase the distance between the
> rich and poor. The rich get richer and the poor get poorer. The govern-
> ment will do anything for money. This is God's land. I am just the ad-
> ministrator for God. . . . My enemies are the people in the government
> who are trying to stop me from distributing the land to the people. Ex-

amples are the Land Regulatory Commission and Bureau of Lands who are making titles that are overlapping. My enemies are the syndicates in the government. What the government is doing is, for example, this is your land, and they will give it to another party so the two of you will be like two roosters in a cockfight.

Monarco claimed that his real estate business did not differ from the commercial transactions of government officials. Pasero, when I asked him if his corporations were a syndicate, had a similar reply: "The word syndicate is nice. The CIA is a syndicate, the Mafia is a syndicate, the Marcos syndicate, the Yacuza syndicate, all are existing so the word is very nice."

THE CORPORATION FACES
GRASSROOTS DEMOCRACY

A commissioner of the Presidential Commission on the Urban Poor explained to me that the syndicate leader Don Pasero had been more successful in providing housing to the poor than any governmental agency. The intergroup network known as Sama Sama also had realized the government's inability to respond to the poor in planning housing or communities. With the group's anchor in the Catholic church, urban poor squatter leaders of Sama Sama were schooled in a view of the Philippines that was more global and theoretical than the syndicate leaders' self-presentations as business persons-politicians. Through seminars and NGOs, Sama Sama learned of imperialism; colonialism; sexism; and contemporary challenges to the poor presented by notions of economic development, seen as a preoccupation of Western, Japanese, and Chinese influences in Southeast Asia. Sama Sama used its vertical intergroup network to seek urban poor Filipino political control of the land they occupied, with the consent of the federal government, while weakening official administrative influences in Payatas of organizations linked to predominant notions of development. The poor would gain a foothold in the capitalist economy through the capital provided by ownership of the land they had been developing as squatters, not through the corporate hierarchies of syndicates, or through demolition of their homes and forced relocation to colonies where, as cheap labor, they could serve the goals of industrialization.

One of Sama Sama's controversial goals was to avert the creation of segregated communities of the poor. Although squatter settlements were often criticized for their not-so-poor residents and absentee landlords (a characteristic of syndicate-controlled land), economically heterogeneous communities provided the very poor with linkages to patrons and employment. In contrast to the bureaucratized and leader-focused syndicates, Sama Sama accomplished its goals, in part, through democratic processes espoused by President Aquino and her governmental appointees. Unlike syndicates and other squatter organizations, at the base of Sama Sama was an elected group of leaders from within the settlement. Through meetings of chapters and their elected leaders, and in close consultation with NGOs, Sama Sama constructed block organizations that were to divide among households the land they occupied as each household purchased its land with the assistance of a loan program within the federal government. Each household would be able to construct (or retain) the house it could afford—from a one-room shanty to a two-story concrete block house with balconies and glass windows (see Parnell 2002).

Sama Sama's strategy was to recruit members who were already allied with the many other associations within Payatas (without defining the groups as mutually exclusive) to create a settlement-wide democratic base from which Sama Sama could argue representation of the majority of National Government Center households. It was to be an alternative to more fractured existing forms of local governance: Payatas was divided into four *barangays* with elected leaders, some of whom were said to sell Payatas land, and many of its homeowner associations were registered with the Securities Exchange Commission. Formally, these associations fell under the jurisdiction of the federal Home Insurance Guarantee Corporation (HIGC); informally, many, such as syndicates and some constituents of Consep, a city-wide "squatter" organization attached to past policies of the Bureau of Forestry Development, acted as self-governing autonomous groups with jurisdiction over land areas.

Sama Sama had strongly opposed the dictatorship of Ferdinand Marcos. When Marcos called a snap presidential election for February 7, 1986, members of Sama Sama rallied behind his opponent, Corazon Aquino, who, while campaigning, visited the National Government Center and told the squatters that if she won the election she would set aside a portion of the NGC for its urban poor residents. When Marcos proclaimed victory, members of Sama Sama joined other Filipinos in protest along Epifanio de los Santos Avenue to form the People Power Revolution, which quickly drove Marcos out of the country and assured Aquino's ascension to the presidency on February 25.

Through Catholic community organizers and church leaders, Sama Sama formed close advisory and planning associations with a prestigious international development firm and Aquino-appointed leaders within federal housing bureaucracies. Sama Sama's goal, in addition to acquiring land, was to form a partnership with the federal government in the planning and development of housing, infrastructure, and communities for the urban poor of the National Government Center. In August 1987, President Aquino, through Presidential Proclamation 137, set aside 150 hectares of the 440 hectare National Government Center for squatters who had resided in the NGC prior to the People Power Revolution of February 1986. The federal government also recognized Sama Sama as the group that would negotiate on behalf of and represent all residents of the NGC. Then, in January 1988, President Aquino, through Memorandum Order No. 151, established the National Government Center Housing Committee and gave Sama Sama, a member of the committee, a vote on policy decisions that was equal to that of all government agencies combined.

When President Aquino set aside a portion of the NGC for its residents and a housing project to be led by Sama Sama, syndicates posted "for sale" signs along the major roadway passing through the NGC. Sama Sama refocused its recruiting activities from the NGC to remaining squatter regions of Payatas, hoping to expand its partnership with the federal government in planning housing and communities. In response, syndicates, Consep, and homeowners' associations from other intergroup networks formed a coalition to oppose Sama Sama and its powerful network. A driving force within the opposition coalition was Samako, (whose members, in the early 1980s, had formed the original base of Sama Sama). Around this time, the Presidential Commission on the Urban Poor and Philippine congressmen began their investigation into syndicates, labeling them criminal. The label became part of the talk within the Sama Sama network, although members of Sama Sama appeared to use it in qualified ways—some, along with their relatives, maintained contacts with syndicate organizations. Also, the syndicates, although openly opposing Sama Sama, continued to provide and protect housing for the poor.

Early in Sama Sama's formation, Samako broke from the mother association as those who became its leaders in the church opposed Marcos and favored the creation of democratic space among the urban poor. Samako, led by the former pro-Marcos vice president of Sama Sama, became a housing collective in search of land. Its leader had openly associated with the National Housing Authority, frequently characterized by Sama Sama as a Marcos-controlled housing bureaucracy that had been notoriously inept in providing housing for the poor and had supported numerous forced relocations and demolition attempts. As Sama Sama grew, the dispute between Samako and Sama Sama intensified, each group recruiting allies throughout Payatas. Then, in 1987, after President Aquino appointed the president of Sama Sama to the planning committee for the NGC, the president of Samako threatened to challenge the existing president of Sama Sama in an upcoming election. Turning democracy against its proponents, the leader of

Samako, if she were to win the election, could wrest Sama Sama and governance of the NGC urban poor out of the hands of the liberal Catholic- and Aquino-aligned intergroup network. At the same time, Samako was forming alliances with syndicate organizations.

Also at this time, the Home Insurance Guarantee Corporation (HIGC) began hearings on a challenge Samako had mounted to an early 1980s Sama Sama election that had just preceded Samako's exit from the Sama Sama network (Samako's president lost that election); in this intervention, the HIGC sought a compromise in the ever-expanding dispute between the two organizations that was threatening the government's hopes for a housing success in the NGC. The two groups refused to compromise, and the hearing became a confrontation over democracy versus the rule of law. The lawyer for Samako, whose president was asking the HIGC to name her the legitimate president of Sama Sama because of alleged fraud committed in the 1984 election, responded to Sama Sama's proposal that the dispute be settled through the upcoming Sama Sama election:

> With deference to *panyero* here, I would like to strongly put on record, very strong objections, your honor, to that proposal for the reason that we have before us a very legal organization [Sama Sama] which was duly registered with the Office given juridical personality and governed by certain charters and by a code of By-Laws. . . . So, we have the Code By-Laws and we have the Articles of Incorporation and therefore, we would insist that all actuations and all activities of this, the members of this organization, should conform with those rules. . . . What the respondent [Sama Sama] would have, however, your honor, is something like people power. . . . We are talking of what happened in 1985, and I would really maintain, your honor, that we should go by the rules because we are governed by rules. (Republic of the Philippines 1987:6–7).

Sama Sama's attorney then compared this case with the People Power Revolution:

> I would like to point out that Sama Sama, which stemmed from the fact it is a people's organization . . . this is an attempt on the part of the people to express their Christianity in a way that they would have to cooperate. Now, if that is true, and that still remains the inspiration of the Sama Sama, I doubt and I [think] if [Sama Sama's opponent] would subject herself to the people's mandate if in case she really represents [the interests of] Sama Sama. This is beyond legality, your honor. . . . This is no case of legality, your honor. For that matter, an analogy can be presented. [The]

February revolution is one example. Marcos, up to now, with due respect to the loyalists, Marcos [residing in Hawai'i at the time] insists that he is still the President of the Philippines, and what is his legal argument? He says that the Batasan Pambasa [Congress] proclaimed him as President and by the will of the people, the sovereign people, should be given priority. That spells the difference between democracy and other forms of government. (ibid.: 18–20).

Samako's attorney responded:

Counsel seems to disregard all rules. What prevents what? We go by rules in this country, and I think we are bound by certain rules so it will not be chaotic; otherwise, we will have people's power every time, and people's power is not really good power, you know. You never know who are the people who really will represent . . . (ibid.: 23).

As this hearing was taking place, and as the PCUP and Congress were continuing their attacks on land syndicates in Payatas, Samako was shifting its network alliances. Not willing to compromise and gain land through Sama Sama's housing project, Samako had shopped around for sponsors, including a possible association with a small energy company in Texas. Eventually, they joined the intergroup network of an aspiring real estate developer venturing into the Payatas region. Although poorly linked to other groups, he had established ties to Don Pasero and, through this association, was promised a portion of Payatas land that included an abandoned uncompleted concrete block subdivision. Opening an office on the land, he moved members of Samako into the empty structures. As I was interviewing the developer in his main office outside Payatas, he received a telephone call notifying him that his office in Payatas had been bombed by the Philippine military (acting on governmental orders). The military then stationed soldiers on the Samako occupied terrain, which Samako eventually abandoned. Samako's leader, as she was preparing to challenge the Sama Sama network in its upcoming presidential election of squatter leadership, then sought a direct alliance with Don Pasero, who promised to locate the organization on land near his home in Payatas.

Following this offensive against the syndicates and Samako, I interviewed a Sama Sama ally who was a leader in federal housing bureaucracies. He explained that through talks with a respected Philippine general he hoped to participate in forming a permanent military unit that would combat syndicate control of land in Metropolitan Manila. I also interviewed

Don Pasero who, when questioned about the armed offensive against his associates in Payatas, provided two explanations: in one, he suggested that the real estate developer who had acted as Samako's patron had failed to pay him for the land in Payatas on which he located Samako's members. In this, Pasero aligned himself with the military offensive. In the other explanation, Pasero repeated rumors that identified a leader of the Congressional hearings against syndicates as a real estate developer in alliance with two powerful Payatas figures who claimed control of the land that Samako had occupied.

Members of Sama Sama did not attempt to aid the members of Samako, although they had, while I was conducting research, used past demolition attempts in Payatas to organize homeowners into Sama Sama allies. Among governmental officials who backed the demolition (and, it appears, participated in its organization) were Sama Sama allies who had fought successfully to gain a moratorium on demolition in the Payatas region. No other demolition attempts had gone forward without their opposition. Nor were other demolitions attempted on syndicate lands during this period. But Samako, like the land syndicates, had weakened its urban poor identity through using the (HIGC) rules rather than compromise and democratic elections in its attempt to gain control of Sama Sama. Samako's association with the syndicates highlighted this strategy and its leader's past association with Marcos cronies in housing bureaucracies. The national campaign against syndicates and Sama Sama's dispute with an uncompromising Samako interlinked as Samako invoked the law against people power while allying with land syndicates. In response to Samako, Sama Sama temporarily abandoned its major cause—to prevent demolition of squatter homes and to defend squatters' rights to live on the land they occupied. Samako's, and the syndicates', pursuit of the same goals was unacknowledged.

Seven years later, when I returned to Payatas, the syndicates were still operating and Samako had dissolved. Sama Sama and its members remained in the NGC and Payatas. They had won the fight against demolition and relocation for all of the residents of the NGC and continued to form block associations for the distribution of land. I am told that Sama Sama continues its work today, as do the syndicates, although it is said that the aging Don Pasero was jailed on arms charges and, eventually, died.

CONCLUSION

The process of criminalization in Payatas Estate symbolized a shift in power across intergroup networks, rather than a breaking of the law. Although from the perspective of official law there was quite a bit of illegality in Payatas, it had rarely been criminalized. Breaking the law was not so much the issue in criminalization of land syndicates as law's relationship to democracy. The syndicates' chosen association with the law that was seen to impede and to have impeded Filipino access to Philippine resources merged with the syndicates' perceived associations with, for example, colonialism and international capitalism, all of which were considered contradictory to Philippine life. As syndicates were seen through these associations by those taking democratic stances, the syndicates' other qualities were elided to produce their condemnation. The iconization of syndicates, which was effective in Manila only from within a sense of democracy as an enlightened form of the state, rendered opaque the syndicates' other formerly heightened qualities as a populist alternative to a state often seen as corrupt and as a provider of residential stability in place of forced relocation. The process of criminalization that followed the People Power Revolution was then also a process of simplification in the context of change as well as an assertion of Philippine rights through the invocation of crime but not law.

That a list produced by the Presidential Commission on the Urban Poor could be so powerful in eliciting criminality where it had not existed, even given the syndicates' past vulnerability to such accusations, suggests the profound nature of change wrought by democratization and the homologenizing and iconizing forces that followed. But it also appears that the process of criminalization that arose in disputes among intergroup networks stretching out from Payatas was never legalized or institutionalized. Perhaps, like many official governmental processes, criminalization lost power in the swirl of urban poor Filipinos pursuing multiple associations. In such a context, the simplification of identity is hard to pull off for long.

ACKNOWLEDGEMENTS

My research in the Philippines was made possible by a Fulbright Senior Research Scholar Award, a Research Leave Supplement from Indiana University, and a President's travel grant from International Programs at Indiana University. While in the Philippines, I enjoyed the collegiality and guidance of my colleagues at the Institute of Philippine Culture, and am especially grateful to Romana de los Reyes, its director at the time. Numerous faculty members at Ateneo de Manila University were generous in their knowledge of life in Manila. I also want to thank Professor Helen Mendoza for her advice and friendship. I am especially grateful to my research assistant, Connie Bascug, for her time, patience, and advice. I learned much from the community organizers of COPE, who were gracious in their hospitality and guidance. I am especially grateful to the people of Commonwealth, who accepted me as a resident in their communities and allowed me to become part of their daily lives.

BIBLIOGRAPHY

Bourdieu, Pierre
1987   The Force of Law: Toward a Sociology of the Juridical Field. The Hastings
       Law Journal 38: 814–853.
Irvine, Judith T. and Susan Gal
2000   Language Ideology and Linguistic Differentiation. *In* Regimes of Language:
       Ideologies, Politics, and Identities. Paul V. Kroskrity, ed. Pp. 35–83. Santa Fe,
       NM: School of American Research Press.
Jackson, Michael, ed.
1996   Things as They Are: New Direction in Phenomenological Anthropology.
       Bloomington: Indiana University Press.
Latour, Bruno
1986   Laboratory Life: The Construction of Scientific Facts. Princeton, NJ: Prince-
       ton University Press.
McCoy, Alfred W. and Ed. C. de Jesus, eds.
1982   Philippine Social History and Local Transformation. Southeast Asian Publi-
       cation Series, 7. Quezon City, Philippines: Ateneo de Manila University
       Press.
Parnell, Philip C.
2002   The Composite State: The Poor and the Nation in Manila. *In* Ethnography
       in Unstable Places: Everyday Lives in Contexts of Dramatic Political Change.
       Carol Greenhouse, Elizabeth Mertz and Kay B. Warren, eds. Pp. 146–177.
       Durham, NC: Duke University Press.
Republic of the Philippines
1987   Transcript. HIGC Case No. 52. Manila: Home Insurance Guarantee Corpo-
       ration.
Wagner, Roy
1981   The Invention of Culture. Chicago: University of Chicago Press.

CHAPTER 9

MAFIA WITHOUT MALFEASANCE, CLANS WITHOUT CRIME

The Criminality Conundrum in Post-Communist Europe

Janine R. Wedel

SINCE THE BREAKUP OF THE SOVIET UNION at the end of 1991, the American stereotype of the "Evil Empire" has been replaced by quite another image of the new Russia: that of "mafia."[1] Western images of the mafia zero in on "criminal" activities,[2] widespread corruption, and their potential threats. As in many stereotypes, that of the new Russia as mafia-influenced holds some truth: Payments to governmental officials are common; contract murders and trafficking in drugs and prostitutes are widespread; the trade in black-market weapons and nuclear materials across borders is widely reported and often presented by Western media and governments as a national security threat. Countering eastern European organized crime and corruption and introducing the "rule of law" has become a growth industry in the United States and some other Western nations.

The wholesale characterization of the societies of eastern Europe as criminal and corrupt by Western analysts, policymakers, and journalists obscures how peoples of the region construct notions of criminality and corruption. From the outside, relationships and practices that eastern Europeans see as benign or at least acceptable may be criminalized. Widely used terms such as mafia, frequently employed to make sense of society in the face of dramatic change, are easily misinterpreted in the West.

Today, eastern European concepts of criminality and corruption grow out of a framework of law, relationships, and mindsets rooted in the previous communist system and shaped by the post-communist years of reform. During the communist years in eastern Europe, views of law became deeply rooted in the dichotomies of communism versus capitalism and state versus society. Under communism, eastern Europeans qualified their relationship to the communist state bureaucracy by creating a distinction between state and society; as people popularly referred to themselves as society, they used informal social networks to oppose and circumvent the state. Now, after the fall of communist governments, these networks of exchange—doing politics and business through and within them—continue to play a role in configuring people's perceptions of law, justice, and crime. Importantly, those networks for getting things done within and outside the state that brought the personalized networks of the citizens into the state economy and bureaucracy, continue to break down barriers between state and private spheres, creating hybrid organizations for getting things done that, from a Western capitalist perspective, consistently violate prescribed roles of the state in the private economy and of private organizations in governmental processes. Especially from the perspective of Western conceptualizations of boundaries between state and private spheres, popular forms of eastern European society can appear criminal.

While the state versus society ideology galvanized eastern Europeans' views, they reorganized the state-society separation through informal social networks. Informal networks blur spheres—state and private, bureaucracy and market, legal and illegal—that, in the West, generally are somewhat more separated and bounded by concepts of crime and corruption. Yet, when Western institutions export the rule of law, they often impose assumptions on eastern European societies that may not coincide with indigenous views of how state and private should interrelate, which are nuanced. For example, anticorruption programs conducted by the World Bank and other international organizations are grounded in a widely employed definition of corruption—"the abuse of public office for private gain" (PREM, The World Bank, 1997:8)—and are often based on idealized notions of state-private relations that may not apply even in the West. What are the meanings and roles of the concept of corruption where the state-private distinction is unclear?

In eastern Europe, the state-private divide may be fluid, subdivided, overlapping, or otherwise obscure. There, skillful actors blend, mediate, and otherwise shape state and private, market and bureaucracy, and legal and illegal for personal and group benefit. From the West, such state-private mixes can easily be misinterpreted as criminal. Yet every activity that crosses the state-private divide is not inherently corrupt.

Moreover, in eastern Europe people do not necessarily equate violating the law with criminality. Westerners may be accustomed to fairly clear standards for judging who is guilty (even if these are not always or even often applied), assuming that criminality elsewhere is determined on the basis of such guidelines; but, in Russia, for example, the standards applied to decide who is criminal may not be as clear. As under communism, the legal prosecution of an alleged perpetrator may depend on factors in the political-economic domain—such as the accuser's political and economic affiliations, current positioning, and economic goals—and the political and economic affiliations and current positioning of the alleged perpetrator. The law, which is less a system for expressing shared ideals than it is a mechanism for exercising power in social relations, can be used as a weapon by one group against another.

Another point of East-West misunderstanding concerns widespread eastern European use of the term mafia. Labeling some individuals and groups in the region as mafia has become common practice. Throughout the region, the allegation of being mafia can be heard in reference to a wide variety of groups and activities, from officials who accept bribes and former *nomenklatura* (communist) managers who acquired state factories at firesale prices to common street criminals and ex-convicts with their own armed police forces. As I have observed in fieldwork in Poland and Russia, *mafia* also can have an ethnic dimension; for example, Poles talk of Russian mafia and Russians of Chechnyan mafia. One ethnic group accuses another of being mafia.

The concept of mafia can only be understood in its context. In eastern Europe, it is a popularly accepted framework for public expression. *Mafia* has come to organize and symbolize the experiences of many people whose societies, shaped by a communist past, have undergone dramatic and sometimes unsettling change. The mafia framework enables people who are fearful and faced with uncertainty to place blame and locate the sources of change. Allegations of mafia seem to appear most acutely in countries undergoing the deepest social and economic upheaval. As Nancy Ries (1998) documents in Russia: "The mafia is . . . present in everyday talk and in popular culture: mafia is a key symbol through which people convey their perceptions and moral evaluations of systematic transformation."

Today's meetings of eastern Europe and the West therefore create a situation in which both Easterners and Westerners may misuse the concept of criminality to refer to the same social phenomenon, but for different reasons. Popular eastern European ways of networking that blur state-private boundaries that are part of Western notions of the state (that have traveled to eastern Europe) may appear illegal and corrupt and, from the West, may be incorrectly conflated with mafia-like practices. In eastern Europe, new inequalities and uncertainties that have accompanied the fall of communism and the arrival of capitalism may lead to eastern Europeans borrowing the term mafia to express their disappointments while viewing as criminal networks that provide influence within the state and access to wealth within the changing economy. In both cases, using criminalizing concepts to identify differences (be they organizational or merely not what was expected) mitigates against understanding how different historical experiences of the state can produce different popular practices in the blending of state and private sectors. To unravel these East-West disconnects, I will examine the contexts in which mafias—real and imaginary—developed in eastern Europe and explore factors that might account for their appeal. I will show how legacies from the communist past shape mafias as public expression and how eastern European crime and corruption are often misinterpreted in the West.

DIRTY TOGETHERNESS I

Finagling is a legacy from the communist past that helps explain why Westerners tend to criminalize eastern European informal systems of exchange. By finagling, I mean the informal exchange relationships that developed under communism to obtain scarce information, resources, services, and privileges (see, for example, Wedel 1986 and 1992). In Russia, this was known as *blat* (see, for example, Ledeneva 1998). The state—its monopolistic economic, political, and legal control—played a crucial role in the evolution of finagling by spawning informal groups and networks that gained influence over eastern European markets and institutions, both under communism and under the reforms of post-communism. Under communism, the key to state power was an expansionist bureaucracy that monopolized the allocation of resources (Verdery 1991, 1996). Economic decisions were made in the political domain, and control over resources insured state power. Demand always outpaced supply, creating economies of shortage (Kornai 1980).

Individuals and groups within state organizations developed informal social networks to circumvent shortages, bureaucracy, and the constraints of central planning. In bypassing official channels, they transformed many state distribution procedures, as documented by anthropologists of Central and Eastern Europe (for example, see Kideckel 1982 and 1993, Sampson 1986, Wedel 1986 and 1992). Further east, patronage networks virtually ran various regions of the Soviet Union (for example, see Ledeneva 1998, Willerton 1992). Although not explicitly institutionalized, these relationships were regularized and exhibited clear patterns.[3] Informal relationships and practices penetrated and stood apart from the state while, at the same time, being circumscribed by it.

To get things done in the state economy and bureaucracy, eastern Europeans personalized relationships within the state, creating a kind of personalized state (Wedel 1986:50–51). People tended to think in terms of who more than what. A typical list of errands in Warsaw consisted of names (rather than institutions or organizations) matched up with tasks: to repair heating, contact Pan (Mr.) Jan; for gasoline, Pan Piotr; for prompt medical attention, Pani (Mrs.) Jadwiga; or, to reserve a place in a kindergarten or university, Pani Antonina. Often, the most important good was information disseminated through informal networks based on trust—information about who, how, and where was the lifeblood of economic and political survival (Wedel 1986:33–117).

Skirting the system became a way of life with its own language, impulses of discretion, and habits of secrecy. Nearly everyone engaged in what Westerners might consider corruption, such as under-the-table deals and payments, simply to survive or to have a somewhat better life. "Dirty togetherness," Adam Podgorecki's (1987) reference to cliquishness and close-knit networks in the context of scarcity and distrust of the state, was endemic to the communist system.

Finagling spawned its own self-referential agency-obscuring vocabulary that could only be conveyed in context and was only roughly translatable. Terms describing informal "arranging" served to mask the nature of the particular matter or transaction at hand and built an expedient ambiguity into the language and activities of everyday life. Additional uses of language further served to obscure the agency of actors vis-à-vis hard-to-come-by goods and services. For example, Poles spoke not of "buying" or "bribing" to obtain things ranging from shoes to an apartment, but rather of having "received" them (Wedel 1986:43).

Underneath the facade of passivity, however, nearly everyone was vulnerable and therefore potentially guilty. As in economic decisions, legal decisions were vested in the monopolistic control of communist authorities. Without standards independent of politics, the law was often applied arbitrarily. Whether an individual was accused of a crime often was based more on his identity rather than distinctive behavior, and the identity of an alleged perpetrator often determined the definition and severity of the crime. Ilona Morzol and Michal Ogorek discuss law's application in communist Poland (1992:62): (Laws were drawn ambiguously and imprecisely of set purpose— the better to apply arbitrarily. One could not rigorously ascertain whether someone was guilty of a given offense or whether a given act was criminal. The whole system was set up so as to make it possible that anyone subject to the system could be convicted or acquitted of one charge or another, at the complete discretion of state power. As a popular saying went: Give me the person, and I'll find the law [that he broke]."

In such a discretionary system, law was an effective means of subverting one's opponent. Criminal charges by one group against another within the power apparatus were a crucial political weapon in the arsenal of Communist Party authorities. For example, a Polish anti-corruption campaign of the early 1980s targeted, in part, former high-ranking communist officials for investigation and/or prosecution for "economic crimes." The martial law government of General Wojciech Jaruzelski had identified these officials as its political opponents (Wedel 1986:51–52). Successful prosecution of a rival group as corrupt or criminal could render it a discredited non-player.

LEGAL PLURALISM

Given state control over the economy and state ownership of property and production in communist eastern Europe, property belonged to *everyone and no one*. What, then, was ownership and on what moral basis could claim be laid to personal, public, and state property? From an individual's point of view, goods belonging to everyone and no one could potentially be acquired by and belong to an individual. Such practices were not considered "stealing" in Polish factories of the 1980s. In an article entitled "When Theft is not Theft," Elzbieta Firlit and Jerzy Chlopecki (1992) detail the nuances of morality among Polish factory workers under communism. A worker's setting aside goods belonging to the state-owned factory—to everyone and no one— to take home for use in his own *side job* was merely *lifting* and, therefore, acting morally. On the other hand, another worker's taking from his follow worker that which had already been set aside for personal use was considered *stealing* and morally wrong. As Firlit and Chlopecki (1992:97) observe:

> On the ethical level, the continuum ranges from what is commonly condemned, at least as a matter of form, to what is openly justified or even acclaimed by public opinion. . . . It [is] necessary to distinguish among theft, lifting, "arranging," doing favors (for no pay), exchanging services, *handel,* "side jobs," and bribery . . . each of these activities has a different social meaning and implications.

In a system in which nearly everyone engaged in dirty togetherness, people developed ethical systems in which legality was seen to diverge greatly from morality. Their experiences of law and morality did not stem from fixed notions of justice and its universal application, such as those that are sometimes articulated by and inscribed on citizens through state systems. As I note in Poland (1986:61): "What is legal is often not considered moral; what is illegal is often considered moral. In thinking about how to obtain quality medical care, acquire tickets for Jazz Jamboree (an annual international jazz festival in Warsaw), or emigrate—whether legally or illegally—people weigh moral and pragmatic concerns, but not legality. In a society in which people find it necessary to slight the system, the boundaries between legal and illegal are understandably fuzzy."

Caroline Humphrey observes a similar view in Russia. She explains (1999:199): "In Russia, perhaps more than in other countries, people who engage in activities defined by the state as illegal do not necessarily define themselves as criminals. Stalin's harsh legal policies, which defined actions such as tardiness at work, aiding abortions, or accidental loss of secret documents as crimes . . . reinforced the long-standing Russian attitude that divorced community from state notions of law (*zakon*)."

Such legal pluralism was compatible with the tendency under communism toward dichotomous patterns of thinking that divided people into mutually exclusive groups. *We* versus *they*—internationally, the communist world versus the capitalist one—took on a powerful domestic variant: we the people (society versus the communists and/or the state. As David Kideckel (1994:141) writes:

> This [communist] social system forced a dichotomous division of society into clearly demarcated public and private spheres characterized by those with absolute power in the former and those who run from power and responsibility in the latter. Thus, living in this social system, East Europeans were trained by experience to divide the world into two mutually exclusive categories of "Us and Them," the unfairly privileged and powerful few and the vast majority of long-suffering, decent folk; party nomenklatura and citizenry.

Today's discretionary use of the law in eastern Europe and Russia is then deeply rooted in the powerful we versus they mindset honed under communism. In a system in which extra-legal factors often determined the outcome of judicial decisions, people came to see communist authorities and the state as the all-powerful Other.

DIRTY TOGETHERNESS II

What happened to these informal systems in 1989, when the communist regimes of Central and Eastern Europe collapsed, and in 1991, when the

Soviet Union broke apart? Theoretically, there were two possibilities. Informal systems could have *supported* the development of new official state institutions; or, they could have *obstructed* such institutional change.

The aftermath of the fall of communism was an "open historical situation"—a period of immense change in which structure is so in flux that it
provides myriad possibilities—as Karl Wittfogel (1981:8, 15ff, 437, 447f)
has described it. In the legal, administrative, political, and economic flux
that followed the collapse of communist governments, many informal
groups, empowered by the erosion of the centralized state and enticed by
myriad new opportunities for gaining access to state resources, took advantage of these opportunities. The open moments following the fall of communism encouraged a free-for-all in which many resources and
opportunities were divvied up. The people who were most energetic and
well-positioned to take advantage of opportunities were the most successful.

Dirty togetherness thrived. Informal systems played a pivotal role in
many reform processes of the 1990s—from privatization and economic restructuring to public administration and the development of the nongovernmental (NGO) sector. Informal systems became integrated with the reforms
themselves and helped shape their development. By providing unrestrained
opportunities for insiders to acquire resources, some reforms fostered the proliferation and entrenchment of informal groups and networks, including
those linked to organized crime. In Russia there was mass *grabitization* of
state-owned enterprises, as many Russians came to call the privatization that
was linked en masse to what Russians called mafia (for example, see Wedel
2001:138–142). The "reforms" were more about wealth confiscation than
wealth creation; and the incentive system encouraged looting, asset stripping,
and capital flight (see for example, Nelson and Kuzes 1994 and 1995, Bivens
and Bernstein 1998, Hedlund 1999, Klebnikov 2000). E. Wayne Merry, former chief political analyst at the U.S. Embassy in Moscow, observed that "We
created a virtual open shop for thievery at a national level and for capital
flight in terms of hundreds of billions of dollars, and the raping of natural resources. . . ."[4] Billionaire oligarchs were created virtually overnight.

Across eastern Europe, groups that coalesced under communism (including *nomenklatura*) have played a major role in shaping property relations
and politics in the post-communist period, as anthropologists and sociologists have documented. In Romania, certain elites—largely members of the
former Communist Party apparatus—work together to control resources.
These *unruly coalitions,* as Katherine Verdery (1996:193) calls them, are
"loose clusterings of elites . . . who are less institutionalized, less visible, less
legitimate" than political parties (1996:194). In Hungary, *restructuring net-*

works shape privatization processes. David Stark (1996; Stark and Bruszt 1998:142–153) identifies the resulting property forms as neither private nor collective, but as "recombinant" property. Stark describes how Hungarian firms develop institutional cross ownership, with managers of several firms acquiring interests in one another's companies. Only people with extensive inside information have the knowledge to participate in such deals.

In Poland, the *srodowisko,* or social circle, helped to organize Polish politics and business during the early 1990s (Wedel 1992:13–14).[5] The circle is dense and multiplex; its members operate in many domains and have multiple functions vis-à-vis one another. Under communism, members of these publicly informal but internally rigorous elite circles worked together for years and developed intricate, efficient, and undeclared networks to survive and even thrive in the face of dangers and difficulties that cemented their bonds. In the post-communist period, members of a few elite social circles have put their fingers into a multiplicity of pies—in government, politics, business, foundations, and nongovernmental and international organizations. Antoni Kaminski and Joanna Kurczewska coined the term "institutional nomads" (1994:132–153) to refer to members of social circles whose primary loyalty is to the circle rather than to the formal institutional positions that members of the circle occupy.

In Russia and Ukraine, sociologists have charted the system of clans (not based on kinship, as in the classic anthropological definition). A clan, as Russian analysts and citizens use the term, is an informal group of elites whose members promote their mutual political, financial, and strategic interests. Olga Kryshtanovskaya (1997) explains:

> A clan is based on informal relations between its members, and has no registered structure. Its members can be dispersed, but have their men everywhere. They are united by a community of views and loyalty to an idea or a leader. . . . But the head of a clan cannot be pensioned off. He has his men everywhere, his influence is dispersed and not always noticeable. Today he can be in the spotlight, and tomorrow he can retreat into the shadow.

My notion of the "clan-state" (Wedel 2001a and 2001b) builds on Thomas Graham's (1995, 1996) observation of clans whose influence can be countered only by competitor clans. Under the clan-state, certain clans are so closely identified with particular ministries or institutional segments of the Russian government that their governmental and clan agendas and activities sometimes seem identical[6]—there is little separation of the clan from the state. The clan is at once the judge, jury, and legislature. As a system of governance, the clan-state lacks outside accountability, visibility, and means of representation for those under its control.

Throughout eastern Europe, informal systems have helped to configure institutions and reform processes of the 1990s. Such systems may in fact have played a greater role in the 1990s than during the previous period. Endre Sik and Barry Wellman (1999:248) argue that there is more "network capital" under post-communism than under communism. They write (1999:250):

> The transition from communism to postcommunism involves growing uncertainty that is manifested by increasing incidents of minor troubles, crises, calamities, opportunities that must be seized instantly, changes in the rules of the game, and new games with new players. People who lack other alternatives tend to use network capital when conditions worsen or uncertainties prevail. Consequently, network capital increasingly is important for households and firms as a means to cope and grab. This is not only because of the inertia of former practices, but because people rationally rely on their already existing behavioral patterns, skills, and heavy investment in network capital. Under postcommunism, both the culture of networking developed during the communist period and investments in network capital are assets that are proving effective for coping with economic troubles and exploiting available opportunities.

INSTITUTIONAL MANIFESTATIONS OF DIRTY TOGETHERNESS

I call organizational structures set up by informal groups to cross-cut and mediate institutional spheres *flex organizations* in recognition of their adaptable, chameleon-like, multipurpose character. Examining these organizations helps to illustrate how popular practices in eastern Europe blur state and private domains. Flex organizations appear to have become institutionalized at high levels of the Polish state. Legislation since the fall of communism has enabled the creation of profit-making bodies variously called foundations or agencies. These bodies make it legally possible for private groups and institutions to appropriate public resources for themselves. Kaminski (1997: 100) maintains that "The real aim of these institutions is to transfer public means to private individuals or organisations or to create funds within the public sector which can then be intercepted by the initiating parties."

A prime example of flex organizations are *agencje* (agencies) that have been created in all Polish ministries with control over property. These include the ministries of transportation, economy, agriculture, treasury, and defense, according to Piotr Kownacki, deputy director of NIK (Supreme

Chamber of Control), Poland's chief auditing body.[7] Formally nongovernmental organizations, agencje are set up by state officials, attached to their ministries or state organizations, and funded by the state budget. The minister typically appoints an agencja's supervisory board; his selections are often based on political connections, according to legal analyst Jan Stefanowicz.[8] Some 10 to 15 percent of an agencja's profits can be allocated to "social" purposes: if the agencja accrues profits, those profits go to the board, sometimes being funneled into political campaigns. On the other hand, any losses are covered by the state budget.[9]

Agencje have several distinguishing features (Kaminski 1997:100). The agencje's unclear functions and responsibilities are a defining characteristic. They are formally nongovernmental but use state resources and rely on the coercive powers of the state administration. They have unclear status. From the government's point of view, the entities are legally private; from the point of view of the entities, they are public institutions. They have broad prerogatives that are supported by administrative sanctions and are subject to limited public accountability.

Agricultural agencje offer a case in point. With so much property under their control, including cooperative farms inherited from the communist past, agencje have begun "to represent [their] own interests, not those of the state," according to NIK Deputy Director Kownacki. He observes that (most of the money is taken by intermediaries (and the state has very little control over this process.[10] Coal mining and arms also are dominated by agencje and present myriad opportunities for corruption, reports Kownacki.[11] Former NIK Director Lech Kaczynski notes that, under the system of agencje, "much tax-payer money flows to private hands on a large scale."[12] The number of agencje is growing.[13]

These entities are enshrouded in ambiguity. They are part and parcel of the "privatization of the functions of the state," as Kownacki puts it, and they represent "areas of the state in which the state is responsible but has no control."[14] It is precisely such ability to equivocate that may afford these entities their strength and may in part explain the potential influence and resilience of the state-private relationships they embody. These flex organizations involve individuals, groups, and institutions whose status is difficult to establish, and they are unlikely to vanish; on the contrary, they appear to be institutionalized.

Some 30 percent of the Polish budget lies somewhere between the private and the state sector, according to Stefanowicz.[15] The net effect of such state-private relationships may be the *enlargement* of the state sphere. Kaminski (1996:4) argues that post-communist legislative initiatives have facilitated "an indirect enlargement of the dominion of the 'state' through founding of institutions that in appearance are private, but in fact are part of the [appropriated] public domain."

While government officials may traverse state and private spheres to achieve their purposes, they also may engage in boundary blurring *within* the state sphere. Alexei Yurchak (2002) notes two separate spheres in the Russian state: the officialized-public and the personalized-public. These spheres represent different types of practices that coexist and can overlap in the same context. Yurchak (1998, 2002) describes a dual state-private structure within a government agency. Russian entrepreneurs, he has observed, seek protection from state organizations whose officials, at the same time, call upon anticrime measures available to them through law and the assistance of criminal affiliates and groups. This underscores the role of law as manipulable. Yurchak (2002:301) explains that the actors involved

> distinguish between those state laws that they perceive as meaningless and counterproductive and those that they perceive as meaningful and important. The former type of laws (e.g., unreasonably high taxes, constraints on the withdrawal of cash from accounts, privileges given to random groups of citizens) they treat as a formality that has to be followed in officialized-public terms only and that, in fact, can be subjected to hybrid entrepreneurial technologies. The latter type of laws they follow in earnest. Perceiving the state and its laws in accordance with this hybrid model means always expecting that some steps and regulations of the state will be positive and meaningful and some will be negative and unreasonable. The entrepreneurs have to relate to the state in this discriminating manner all the time.

This way of operating need not be cynical. As Yurchak (2002:302) argues, it "allows entrepreneurs to be involved in informal activities and at the same time have a genuine desire for the democratic rule of law in the country."

BLURRED BOUNDARIES
AND THE PERCEPTION OF CRIME

Among those who view eastern Europe from the West, there is a tendency to allege corruption and "conflict-of-interest" without examining the complexities of eastern European relationships and how they might affect constructions of crime and corruption. Underlying efforts by Western governments and international organizations to combat corruption and encourage rule of law are conventional vocabularies and models that infuse Western public administration, comparative political science, sociology, popular discourse, and policymaking. The tendency to uncritically apply these vocabularies and models to eastern Europe may contribute to the misunderstanding of eastern European relationships and their conceptualization as illicit or criminal. The informal groups and networks described above cannot be accurately understood in terms of these vocabularies and models, which may be insufficient to probe changing state-private and political-administrative relations in any complex state.

Three properties common to these groups and networks show their incongruity with Western conceptions and help explain Western misperceptions of eastern European crime and corruption. The first property of these informal systems is that *the unit of decision making is the informal group*. As Westerners look to those capitalizing eastern European nations they tend to criminalize and overemphasize the role of individuals without a sense that individuals are acting as part of a group whose members' agendas and activities are interdependent. Because a network grouping such as a clan is a different unit of economic analysis than is usually considered by economist analysts, who tend to think of individuals as the primary unit to take advantage of economic opportunities, Western analysts tend to blame individuals rather than groups in eastern Europe for violating Western institutional boundaries.

In the contexts of uncertainty and weakly established rule of law, individuals must take the interests of their groups into account when making choices about how to respond to new opportunities. Operating as part of a strategic alliance enables members of the informal group to survive and thrive in uncertainty and indeterminacy. Informal groups and networks have access to state resources through various members, and they maximize their flexibility and influence precisely by blending and traversing different spheres and domains. In the cases of both Polish institutional nomads and Russian clans, a civil servant (dependent on the tenure of a specific political leadership, if not actually brought in or bought off by it) is typically more loyal to his or her group than to an official office or position. In both cases, resources and decision making in economic, political, and societal domains tend to be concentrated in just a few hands.

The second property of these informal systems is that *informal groups and networks operate in, mediate, and blur different spheres*—state and private, bureaucracy and market, legal and illegal—that are well-bounded in the rhetoric of Western public administration. The group's strength derives in significant part from its ability to access the resources and advantages in one sphere for use in another. Informal groups derive influence from their ability to bridge categories and penetrate institutions that are officially separate.

The widely used definition of corruption—"the abuse of public office for private gain"—presented earlier, reflects a dichotomous way of thinking that separates the public and private in ways that are often not applicable in eastern Europe. Ken Jowitt (1983:293) holds that "a major weakness of current approaches to corruption" is that they emphasize "the difference between public and private aspects of social organization" which "makes it impossible to specify the existence and meaning of corruption in settings where no public-private distinction exists institutionally."

POLITICIZED LAW AND THE PERCEPTION OF CRIME

The third property of these informal systems is that *informal groups and networks operate in the multiple domains of politics, economics, and law.* The continued interdependence of the domains of politics, economics, and law and the use of one domain to extract advantages in or leverage another are part and parcel of the arbitrary application of criminality. Access and success in one domain are often contingent on access and success in another. Informal groups and networks can wield influence and control resources to the extent they do because of the nature of the legal contexts in which they operate. To varying degrees, as one informant put it, "the rules are what you make them."

Informal groups and networks in the region evolved (or continued their evolution) in a context in which the communist state's monopoly control over resources was crumbling or had collapsed and opportunities for filling the void abounded. Although the Communist Party ceased to exert monopoly control over the economy, ability to access economic opportunities often remained contingent on political connections. Terms such as "oligarchs"[16] and "financial-industrial groups," which are now widely employed to describe the structure of power and the wielders of influence in Russia, capture this quality of interdependence among domains. Virginie Coulloudon (1998:545) describes this in Russia:

> Lack of transparency is perhaps what differentiates most Russian elite groups from Western lobbies. In contemporary Russia, it is still impossible to make one's money yield a profit without negotiations at some point with state agents. Financiers, industry managers, journalists and scholars agree that one's career depends on one's ability to weave political networks. The constant struggle between elite groups to appoint their protégés to strategic posts does not challenge the legitimacy of the state. On the contrary, it helps to strengthen it.

In Russia, the political-economic structure that has evolved under post-communism differs from communism in two major respects. First, no single group allocates resources, as under communism, although a single group can monopolize an entire sector or sectors (for example, Russian aluminum or gas). Second, the relationship of power to property is no longer one way. As Graham (1999:329) expresses it, "Not only can power be converted into property; property can be converted into power."

For its part, the use of law under post-communism, as under communism, remains highly discretionary and is highly compatible with the political-economic structure. Breaking the law is not necessarily what determines criminality because many people, in different walks of life, routinely violate the law. The law is often used for ad hoc purposes, bargaining, and extracting advantages. Mafia groups have been known to turn over to the police information that incriminates rival groups. As Alena Ledeneva (2001:13) writes:

> *Anybody can be framed and found guilty of some violation of the formal rules,* because the economy operates in such a way that everyone is bound to be involved in some misdemeanor. For example, everybody is forced to earn in the informal economy in order to survive—a practice that is punishable, or could be made so. Businesses are taxed at a rate that forces them to evade taxes in order to do well. Practices such as the embezzlement of state property or tax evasion become pervasive. Inside state institutions, a whole family of corrupt practices, such as bribe-taking and extortion in the granting of licenses, has been prevalent. The fairly ubiquitous character of such practices makes it impossible to punish everyone. *Due to the pervasiveness of the offence, punishment is bound to occur selectively,* on the basis of criteria developed outside the legal domain.

As under communism, law in Russia is sometimes used to disadvantage or discredit political opponents. As a journalist based in Moscow and St. Petersburg in the latter 1990s reported (Whitmore 2000):

> Today, corruption allegations are dragged out for a number of reasons, and none of them have anything to do with fighting corruption. In some cases, they are a means to rein in or intimidate opponents of the state. This summer's detention of media magnate Vladimir Gusinsky, whose NTV television station had relentlessly attacked [Russian President Vladimir] Putin, was a case in point. In other cases, corruption scandals are initiated by financial clans using friendly (privatized) police and prosecutors against their foes. The message here is simple: If you are loyal, steal as much as you like. If you aren't then watch it!

THE LEGACIES OF COMMUNISM,
THE INTRODUCTION OF CAPITALISM,
AND THE PERCEPTION OF CRIME

The legacies of finagling and the Other supplied the building blocks of the mafias—real and imaginary—that the reforms of the past decade have helped to configure. As mafias have grown, other legacies of communism have provided the vantage points from which citizens view the arrival of global capitalism and assess its realities against their expectations—legacies of relative income equality and little crime accompanied by suspicion of the official state. As in other countries where rapid change, democratic ideologies, and capitalization overlay yesterday's society and experience—such as a South Africa in the throws of millennial capitalism described by John and Jean Comaroff (1999)—the future as the present has not lived up to expectations; in countries such as Russia and Ukraine, a huge divide has developed between a tiny minority with enormous wealth and the vast majority of the population with very little by comparison.

The result, as in South Africa, is moral panic and a search for sources of unexpected inequalities and the future's new corruption. Influenced by Western media, people of eastern Europe have invoked criminal frameworks to articulate and locate the sources of their disappointments and disenfranchisement in popular organizational forms and practices. In an ironic twist, *mafia* has become a symbolic scapegoat in the transition to global capitalism as the people turn their own non-state sources of democracy and economic survival into witch-like social engines of immorality. Crime is associated with getting and keeping resources, newfound huge disparities in wealth, and the fact that such disparities are often much more ostentatious than previously acceptable. All of this fuels the belief that people with privilege have achieved it through dubious, dirty togetherness at the expense of those less fortunate.

A component of the *legacy of relative income equality* is that under communism the boss earned little more than the secretary. Under postcommunism, the biggest bosses—dollar billionaires—have stashed much of their cash in Swiss and offshore bank accounts. The International Monetary Fund estimates that from 1995 to 1999 alone capital flight from Russia exceeded $65 billion (Lopez-Claros and Zadornov 2002:109). The privatization of state-owned resources and economic reforms have facilitated the acquisition of staggering fortunes, especially in countries of the former Soviet Union such as Russia and Ukraine. The result is that the few who were well-positioned to take advantage of the changes often have fared very well, while many others have not.

The Russian population, for example, has suffered increasing hardship during the reform years. One authoritative study determined that 38 percent of the population was living in poverty at the close of the first quarter of 1999, as compared with 28 percent one year earlier. Real incomes in June 1999 were 77 percent of the June 1998 level.[17] Further, Russian citizens became poorer in 1999.[18] At the turn of the millennium, an estimated 70 percent of Russians lived below or just above the poverty line. Still, many Russians and other peoples of the region continued to aspire toward more equitable distribution of wealth.

During the years of reform, another powerful idea galvanized eastern European societies: the idea that people themselves could take advantage of new economic opportunities and accumulate vast wealth. For most people, however, "opportunities" such as pyramid schemes turned into a cruel hoax. Comaroff and Comaroff (1999:293), writing about South Africa, identify a very similar circumstance—"a world in which the *possibility* of rapid enrichment, of amassing a fortune by largely invisible methods, is always palpably present." As they explain (1999:284, 293–294):

> On the one hand is a perception, authenticated by glimpses of the vast wealth that passes through most postcolonial societies and into the hands of a few of their citizens: that the mysterious mechanisms of the market hold the key to hitherto unimaginable riches; to capital amassed by the ever more rapid, often immaterial flow of value across time and space, and into the intersecting sites where the local meets the global. On the other hand is the dawning sense of chill desperation attendant on being left out of the promise of prosperity, that *everyone* would be set free to speculate and accumulate, to consume, and to indulge repressed desires. But, for many, the millennial moment has passed without palpable payback.

The millennial moment, both in South Africa and eastern Europe, is associated with dramatic system change and its ripple effects, as well as larger dynamics of the fall of communism to global capitalism. Comaroff and Comaroff elaborate:

> The rise of occult economies in postcolonial, postrevolutionary societies, be they in Europe or Africa, seems overdetermined. For one thing, these tend to be societies in which an optimistic faith in free enterprise encounters, for the first time, the realities of neoliberal economics: of unpredictable shifts in sites of production and the demand for labor; of the acute difficulties inherent in exercising stable control over space, time, or the flow of money; of an equivocal role for the state; of an end to old political alignments,

without any clear lines, beyond pure interest, along which new ones take
shape; of uncertainty surrounding the proper nature of civil society and the
(post?)modern subject. Such are the corollaries of the rise of millennial capitalism as they are felt in much of the contemporary world.

In eastern Europe, people's shock and difficulty in adjusting to dramatic,
inexplicable, rapid and mind-boggling change following the mostly stable
years of the post–World War II era provide the context for the power of mafia
to symbolically organize and explain people's experience. People ascribe to
mafia the good fortune of others and their own lack thereof. Citizens ask,
"How can it be that *they* have done so well, while *I* am barely surviving?"

Another legacy from the communist experience related to the newfound fortunes of a few is *the legacy of little crime.* Citizens often associate the
acquisition and maintenance of wealth with the growth in crime. Visible,
dangerous, violent, and sometimes organized crime, such as contract killings
of bankers and politicians, has exploded in societies with little prior experience of such crime and very low crime rates but some exposure to Western
stereotypes of the mafia in television and movies. This leads people to invoke
associations of mafia, which one might expect to be behind these crimes. Although much mafia-associated crime is limited to turf battles among rival
groups, average citizens feel a sense of danger. They may become unwitting
victims of violence, even if not its intended targets.

The above two legacies of relative income inequality and little crime are
closely connected to the *legacy of suspicion.* A great deal of suspicion accompanied systemic finagling. Because state propaganda under communism was
untrustworthy and contradicted by everyday life, eastern Europeans learned
to "live in the lie," as Václav Havel (1985) described it—to doubt official explanations. Because so much had to be "arranged" under the table in
economies of shortage, many transactions were shrouded in secrecy. Everyday life required considerable political skill and trust. Who was doing and
getting what and people's real motivations and loyalties were often not what
they appeared. This led to seemingly interminable speculation and suspicion
at all levels of society—from an academic's or bureaucrat's interpretation of
her colleague's promotion to a citizen's explanation for his neighbor's good
fortune. Such uncertainty gave rise to an attempt to control the sources of
uncertainty through invoking the notion of mafia.

Verdery (1996:220) equates this with witchcraft. "Talk of mafia is like
talk of witchcraft," she writes. "[It is] a way of attributing difficult social
problems to malevolent and unseen forces." This appears similar to the South
African focus on witchcraft, in which people accuse others of witchcraft—
literally. Occult mechanisms, write Comaroff and Comaroff (1999:284),

which "have become the object of jealousy and envy and evil dealings" have it that "arcane forces are intervening in the production of value, diverting its flow for selfish purposes." Witches are durable, they suggest, because they "distill complex material and social processes into comprehensible human motives, then, they tend to figure in narratives that tie translocal processes to local events, that map translocal scenes onto local landscapes, that translate translocal discourses into local vocabularies of cause and effect" (Comaroff and Comaroff 1999:286).

It is precisely such thinking in the eastern European context that appears to encourage people to interpret life's vicissitudes today in terms of the influence of mafia. Against the background of uncertainty, economic decline, and a world "in which the majority are kept poor by the mystical machinations of the few," as Comaroff and Comaroff (1999:293) put it, talk of mafia expresses a sense that sinister forces beyond people's grasp are pulling the strings and are to be blamed for their misfortunes. With the label mafia, one points the finger at a certain person or group such as business competitors or political opposition and suggests they are under the spell of sinister powers.

CONCLUSION

In eastern Europe and Russia, the deep-seated ideological divides inherited from the East versus the West and communism versus capitalism oppositions of the Cold War are today being replicated in the form of criminality as Westerners critique and attempt to shape emerging forms of the state in eastern Europe, and as the citizens of those states respond to new divisions—those who are being left out versus those who have made out with the arrival of free market ideologies. In eastern Europe, mafia as an accusation of criminality and immorality is not a throwback to tradition. To the contrary, talk of mafia is a response to people's dissatisfaction with their current, sometimes unhappy life circumstances. It is a way for some who have been excluded from the expected fruits of capitalism to assign blame to those who have harvested those fruits. Talk of mafia in eastern Europe, like talk of witchcraft accusations in South Africa today, is a response to the marginalizations that "millennial capitalism" has dealt many of its recipients (Comaroff and Comaroff 1999:283). Being labeled mafia, like the marginalizing forces of witchcraft accusations against those who have inexplicably escaped harm, is an irrefutable indictment.

The consequences of this may be a uniting of disenfranchised eastern Europeans and Russians with Westerners pursuing their own political and economic agendas in demonization of popular networking practices that, while blurring state and private boundaries, serve as sources of economic participation in the new postcommunist societies. It would be ironic for this to happen through the processes of criminalization that accompany Western rule of law perspectives. The result could be the weakening of primary eastern European counters to centralization of political and economic control in the hands of a few local and international elite players.

Researchers and policymakers who design aid programs for eastern Europe and strategies to combat corruption and criminality there have paid too little heed to ways in which economic, political, and legal domains are interdependent as well as the ways in which the spheres of state and private, legal and illegal are crossed. The export of Western economic and political ways to eastern Europe and Russia may also include the export of criminalization, which could weaken local infrastructures for survival and reduce the benefits of economic and political participation. Anthropologists, by using ethnography to place crime and criminality in context, can help foresee and forestall such devastating consequences of using crime and criminality as components of foreign relations.

NOTES

1. Peter Schneider made this point as a discussant during a panel on "Networking with a Vengeance: Clans and Mafia in Eastern Europe and the Former Soviet Union," organized by Janine R. Wedel for the 96th Annual Meeting of the American Anthropological Association, November 20, 1997.
2. Rawlinson (1998) discusses the graphic and misleading coverage of Russian mafia in the Western press.
3. For further analysis of such relationships, see Wedel (1992), Introduction.
4. "*Frontline* Return of the Czar" interview with E. Wayne Merry, PBS website, www.pbs.org\wgbh\pages\frontline\shows\yeltsin\interviews\merry.html.
5. Many members of the various post-communist governments belong to previously existing and identifiable social circles. For example, while leaders of the first post-communist government of Tadeusz Mazowiecki largely hail from a Krakow Catholic intelligentsia circle, those of the government of Jan Krzysztof Bielecki come from a Gdansk circle.
6. For example, the Chubais Clan, which monopolized Russian economic reform and foreign aid during the 1990s, was closely identified with segments of government concerned with privatization and the economy. Competing clans had equivalent ties with other governmental organizations such as the "power ministries" (the Ministries of Defense and Internal Affairs, and the Security Services). For details, see Wedel (2001a:123–174).
7. Interview with Piotr Kownacki, deputy director of NIK, July 26, 1999.

8. Interviews with Jan Stefanowicz, July 14 and 15, 1999.
9. Ibid.
10. Interview with Piotr Kownacki, Deputy Director of NIK, July 26, 1999.
11. Ibid.
12. Interview with Lech Kaczynski, July 14, 1999.
13. Interviews with Jan Stefanowicz, July 14 and 15, 1999.
14. Interview with Piotr Kownacki, deputy director of NIK, July 26, 1999.
15. Interviews with Jan Stefanowicz, July 14 and 15, 1999.
16. Oligarchy, in its classic definition, means rule by a few, and often accumulation of wealth by a small group that could not maintain power without military and governmental support. See *The* Encyclopaedia of the Social Sciences, 1937.
17. OECD *Economic Outlook,* December 1999: 132.
18. This is the case even though wage arrears and absolute numbers below the poverty line in 1999 trended down. "The average level of Russians' real cash income—incomes adjusted to account for inflation—decreased 15 percent," according to the Russian Statistics Agency's Yevgenia Borisova, in "Poverty Still Widespread Despite Modest Growth," *Moscow Times,* January 13, 2000; also in *Johnson's Russia List,* no. 4032, January 13, 2000.

BIBLIOGRAPHY

Bivens, Matt and Jonas Bernstein
1998 The Russia You Never Met. Demokratizatsiya: The Journal of Post-Soviet Democratization 6(4): 613–647.
Boissevain, Jeremy
1974 Friends of Friends: Networks, Manipulators and Coalitions. Oxford: Basil Blackwell.
Comaroff, Jean and John L. Comaroff
1999 Occult Economies and the Violence of Abstraction: Notes from the South African Postcolony. American Ethnologist 26(2): 279–303.
Coulloudon, Virginie
1997 The Criminalization of Russia's Political Elite. East European Constitutional Review 6(4): 73–78.
1998 Elite Groups. Demokratizatsiya: The Journal of Post-Soviet Democratization 6(3): 535–549.
Encyclopaedia of the Social Sciences, 1937, Edwin R. A. Seligman, main ed. Pp. 462–464. New York: Macmillan.
Firlit, Elzbieta and Jerzy Chlopecki
1992 When Theft is Not Theft. *In* The Unplanned Society: Poland During and After Communism. Janine Wedel, ed. Pp. 95–109. New York: Columbia University Press.

Graham, Thomas E.
1995   The New Russian Regime. Nezavisimaya Gazeta. November 23.
1996   Russia's New Non—Democrats. Harper's Magazine 292 (1751): 26–28.
1999   From Oligarchy to Oligarchy: The Structure of Russia's Ruling Elite. Demokratizatsiya: The Journal of Post-Soviet Democratization 7(3): 325–340.
Havel, Václav
1985   The Power of the Powerless. *In* The Power of the Powerless: Citizens Against the State in Central-Eastern Europe. John Keane, ed. Pp. 23–96. Armonk, New York: M. E. Sharpe.
Hedlund, Stefan
1999   Russia's "Market" Economy: A Bad Case of Predatory Capitalism. London: UCL Press Limited.
Humphrey, Caroline
1999   Russian Protection Rackets and the Appropriation of Law and Order. *In* States and Illegal Practices. Josiah McC. Heyman, ed. Pp. 199–232. New York: Berg.
Humphrey, Caroline and Stephen Hugh-Jones
1992   Barter, Exchange, and Value: An Anthropological Approach. Cambridge, England: Cambridge University Press.
Jowitt, Kenneth
1983   Soviet Neotraditionalism: The Political Corruption of a Leninist Regime. Soviet Studies: A Quarterly Journal on the USSR and Eastern Europe 35(3): 275–297.
Kaminski, Antoni Z.
1996   The New Polish Regime and the Specter of Economic Corruption. Summary of paper to be presented at the Woodrow Wilson International Center for Scholars, Princeton, April 3.
1997   Corruption under the Post-Communist Transformation: The Case of Poland. Polish Sociological Review 2(II8): 91–117.
Kaminski, Antoni Z. and Joanna Kurczewska
1994   Main Actors of Transformation: The Nomadic Elites. *In* The General Outlines of Transformation. Eric Allardt and W. Wesolowski, eds. Pp. 132–153. Warszawa: IFIS PAN Publishing.
Kideckel, David
1982   The Socialist Transformation of Agriculture in a Romanian Commune, 1945–1962. American Ethnologist 9(2): 320–40.
1993   The Solitude of Collectivism: Romanian Villagers to the Revolution and Beyond. Ithaca, New York: Cornell University Press.
1994   Us and Them: Concepts of East and West in the East European Transition. *In* Cultural Dilemmas of Post-Communist Societies. Aldona Jawlowska and Marian Kempny, eds. Pp. 134–144. Warsaw, Poland: IFIS Publishers.
Klebnikov, Paul
2000   Godfather of the Kremlin: Boris Berezovsky and the Looting of Russia. New York: Harcourt, Inc.

Kornai, János
1980   Economics of Shortage. Amsterdam: North-Holland.
Kryshtanovskaya, Olga
1997   The Real Masters of Russia. *In* RIA Novosti Argumenty i Fakty No. 21, May. Reprinted in Johnson's Russia List.
Ledeneva, Alena V.
1998   Russia's Economy of Favours: *Blat,* Networking and Informal Exchange. Cambridge, England: Cambridge University Press.
2001   Unwritten Rules: How Russia Really Works. London: Centre for European Reform.
Lopez-Claros, Augusto and Mikhail M. Zadornov
2002   Economic Reforms: Steady as She Goes. The Washington Quarterly 25(1): 105–116.
Morzol, Ilona and Michal Ogorek.
1992   Shadow Justice. *In* The Unplanned Society: Poland During and After Communism. Janine R. Wedel, ed. Pp. 62–77. New York: Columbia University Press.
Nelson, Lynn D. and Irina Y. Kuzes
1994   Property to the People: The Struggle for Radical Economic Reform in Russia. Armonk, New York: M. E. Sharpe.
1995   Radical Reform in Yeltsin's Russia: Political, Economic, and Social Dimensions. Armonk, New York: M. E. Sharpe.
Podgorecki, Adam
1987   Polish Society: A Sociological Analysis. Praxis 7(1): 57–78.
PREM, Washington, D.C.: The World Bank.
1997   Helping Countries Combat Corruption: The Role of the World Bank. The World Bank: Poverty Reduction and Economic Management.
Rawlinson, Paddy
1998   Reflections of Russian Organized Crime: Mafia, Media, and Myth. Paper presented at the 97th Annual Meeting of the American Anthropological Association, Philadelphia, December 4.
Ries, Nancy
1998 The Many Faces of the Mob: Mafia as Symbol in Postsocialist Russia. Paper presented at the 97th Annual Meeting of the American Anthropological Association, Philadelphia, December 4.
Sampson, Steven
1986   The Informal Sector in Eastern Europe. Telos 66 (winter): 44–66.
Schneider, Jane and Peter Schneider
1994   Mafia, Antimafia, and Question of Sicilian Culture. Politics and Society 22(2): 237–258.
Sik, Endre and Barry Wellman
1999   Network Capital in Capitalist, Communist, and Post-communist Countries. *In* Networks in the Global Village: Life in Contemporary Communities. Barry Wellman, ed. Pp. 225–277. Boulder, CO: Westview Press.

Stark, David
1996   Recombinant Property in East European Capitalism. American Journal of
       Sociology 101(4): 993–1027.
Stark, David and Laszlo Bruszt
1998   Postsocialist Pathways: Transforming Politics and Property in East Central
       Europe. Cambridge, England: Cambridge University Press.
Varese, Federico
1994   Is Sicily the Future or Russia? Private Protection and the Rise of the Russian
       Mafia. Archives Europeennes de Sociologie 35(2): 224–258.
Verdery, Katherine
1991   Theorizing Socialism: A Prologue to the "Transition." American Ethnologist
       18(3): 419–439.
1996   What Was Socialism, And What Comes Next? Princeton: Princeton Univer-
       sity Press.
Wedel, Janine R.
1986   The Private Poland: An Anthropologist's Look at Everyday Life. New York:
       Facts on File.
Wedel, Janine R., ed.
1992   The Unplanned Society: Poland During and After Communism. Columbia:
       Columbia University Press.
2001a Collision and Collusion: The Strange Case of Western Aid to Eastern Eu-
       rope. New York: Palgrave.
2001b "State" and "Private:" Up Against the Organizational Realities of Central and
       Eastern Europe and the Former Soviet Union. Paper prepared for the Na-
       tional Council for Eurasian and East European Research and the National In-
       stitute of Justice, Washington, December.
Whitmore, Brian
2000   Might Makes Right. Transitions Online, October 2. Reprinted in Johnson's
       Russia List #4555, Oct. 3.
Willerton, John P.
1992   Patronage and Politics in the USSR. New York: Cambridge University Press.
Wittfogel, Karl A.
1981   Oriental Despotism: A Comparative Study of Total Power. New York: Vin-
       tage Books.
Yurchak, Alexei
1998   Mafia, the State, and the New Russian Business. Paper presented at the An-
       nual Meeting of the American Anthropological Association, Philadelphia,
       December 4.
2002   "Entrepreneurial Governmentality in Postsocialist Russia," In The New En-
       trepreneurs of Europe and Asia. Victoria Bonnell and Thomas Gold, eds. Pp.
       278–317. Armonk, New York: M. E. Sharpe.

CHAPTER 10

HEAR NO EVIL, READ NO EVIL, WRITE NO EVIL

Inscriptions of French World War Two Collaborationism

Vera Mark

THE GROUP OF THE THREE MONKEYS—one of which holds its hands over its eyes, another over its ears, and a third over its mouth—provides a key metaphor for framing an ethnography of French World War II collaborationism. A verbal counterpart to the group of monkeys, namely the moral coda taken from a late-fourteenth-century French ballad—"*pour vivre en paix, il faut être aveugle, sourd et muet*" ("to live in peace, one must be blind, deaf and dumb") (Taylor 1996)—provides a folk view on how best to insure societal harmony. It also names three of the principal modes of ethnographic perception—the visual, the auditory, and the spoken. In this chapter I detail the ways in which these modes of ethnographic perception worked for and against my understanding of the political and legal past of a man from a small town in southwestern France.

Jacques Dupont (1894–1984)[1] was a cobbler and small shopkeeper for his entire working career. A veteran of World War I, during World War II he supported the politics of Philippe Pétain, head of the French government in Vichy. A member first of the French Legion, Dupont subsequently joined local branches of the *Service d'Ordre des Légionnaires* (SOL) and then the Militia, increasingly extremist organizations, the last openly collaborationist with the occupying Nazi forces (Gordon 1980; Golsan 1996). All three groups were recruited from active and retired military men, including World War I veterans such as Dupont. Dupont was arrested twice during the World War II years. His first arrest occurred a few days after the Allied landing, in June 1944, when Resistance groups went after known Militia members and conflicts between these groups intensified, especially at the local level, in small towns and villages throughout France. Dupont was released after the head of the town's Resistance intervened personally on his behalf. The cobbler's second arrest took place a year and half later, prior to his being tried for collaborationism in a civil court, for which he was judged guilty and sentenced to five years in prison. Released early for good behavior, Dupont was eventually granted amnesty, the date, type, and consequences of which I only learned after a long and circuitous research process.

In the spirit of many of the other articles in this collection, I underscore the fact that I did not intend to study collaborationism in rural France as part of my initial anthropological field research. At the time (the early 1980s), a background in linguistics and folklore led me to pursue formal questions of genre and regional language maintenance. I was drawn to the rich trove of verbal texts and material artifacts created by Dupont, a traditional artisan. In his cobbler's workshop he read aloud to me many of his poems, composed in both French and in Gascon, the regional language, and showed me the folk art objects of his bricolage. However, collaborationism and its aftermath, a key element of the twentieth century French political unconscious, surfaced continuously throughout my research year in my regular conversations with the cobbler and with many other town residents.

"Collaborationism" is defined by French historian Bertram Gordon as the ideological acceptance of fascism, in contrast to "collaboration." In the French political context, both terms have the charged, negative meaning of supporting the Nazis during their four-year occupation of France, from 1940 to 1944. Collaborators trafficked with the Germans solely for material gain or personal advancement, whereas collaborationists possessed a rudimentary political ideology and viewed themselves as partisans of Franco-German reconciliation and the Nazi European New Order (Gordon 1980:17–19; afterword in Golsan 1996). Thus collaboration may refer to ideological, economic, and political cooperation with the Nazis, but also to French citizens' joining of home-grown fascist organizations, most notably the civilian paramilitary Militia, which worked openly with the Germans in the most repressive phase of the war, beginning in 1943 and especially in 1944. Many scholars distinguish "active" collaboration—which involved denouncing, arresting, and deporting both Jews (French-born, foreign-born, and those in transit) and Resistors, including communists and Masons—from the more "passive" collaboration of going along with events at hand.

The Occupation was *the* traumatic period of reference for the people I interviewed, most of whom were born in the first decades of the twentieth-century. The war split apart generations, families, and communities and brought the destruction of a quarter of the Jews of France. The immediate post-war years saw numerous trials and reprisals, and the consequences of some are still felt, more than 50 years later. While the World War II period in France has been studied above all by historians over the past 25 years, and widely represented in literature and film, it has been less studied by anthropologists. Spanning the last 20 years, my own research about the actual events that took place during World War II in the cobbler's home town, and his role within them, has been fraught with tension. In a standard anthropological paradox, the ethnographer wants to make public what the community and a family want to keep private and silenced. This chapter reflects on what it means to conduct research on criminal events for which there may be little or no material evidence, how different forms of writing reconstitute the past and act upon ethnographic understandings of political and social history, and how the French legal system's protection of the "private domain" of individuals' lives significantly impacts research strategies and outcomes.

SEEING, HEARING, SPEAKING, THEN READING AND WRITING ABOUT COLLABORATIONISM

In the fall of 1995 I filed a request to have special permission to consult Jacques Dupont's trial records, located in a departmental archival service in southwestern France. Although my request was eventually granted, following national policy in place regarding consultation of this particular type of document, I had to sign the following statement: "I promise formally on my word not to divulge (neither orally, nor in writing or by photocopy) any names or information with names attached susceptible of harming the persons cited even deceased or their descendants, and to make no photocopy."

Up to this point, my narrative about the cobbler's political past has presented events straightforwardly with the knowledge of hindsight. However, the actual process that accompanied my understanding of events was far less linear. I would now like to go back in time, to the year 1981–1982, which I spent conducting linguistic research on regional language use. Having briefly met the cobbler on a previous research stay in the southwest of France, I settled into his hometown and visited with Dupont daily in his workshop, where I talked with him as he repaired shoes and leather goods. With some shyness and hesitation, apologizing for their simplicity, he began to show me his poems, typically scribbled in French and Gascon on old wrapping papers. The most striking text was "*Mon temps passat*,"[2] a 186-line poem composed

in Gascon about his childhood summers spent working as a cowherd and
shepherd on local farms. In the course of my research year I saw the cobbler
read parts of this poem to visitors to his store. His other texts did not circulate, however, or did so at the very limited level of being sent only to the local
resident(s) who had inspired their composition in the first place.

Just three months after our first meeting, during one of our Gascon-lan-
guage sessions in mid-November, around Armistice Day, Jacques Dupont
shared the most important historical documents of his life with me, namely
the testimonies from fellow town residents who had defended him at his
trial. I downplayed their importance, skirted the issue of the war, and
avoided history, deferring to my main subject of research at the time, re-
gional/national language maintenance and shift. My retreat into regional
language study—an academic exercise focusing on linguistic survivals—
would provide at first a screen from, but later an entry into, history. I could
not identify with collaborationism on scholarly grounds, for as a non-traditional subject in anthropology at the time, its place was in the background,
which would engender certain blindnesses on my part in subsequent years
of research. The family taboo against evoking the war years—which was
shared by many townspeople and operated at the highest levels of the French
state[3]—was repeatedly underscored by the cobbler's daughter, with whom I
was becoming close. I did not want to jeopardize our relationship, for she
was a key gatekeeper to the community's residents, many of whom she in-
troduced to me as they came to her store. It was unclear to me how residents
would react to my work as an ethnographer if I decided to pursue what remained a highly sensitive subject of social pollution. Although I had a visceral rejection of collaborationism, I could sense its power in this
community more than 40 years after the war; it would eventually define my
entire project in follow-up research. Tapping into my deepest emotions and
those of the cobbler's family, weaving between the conscious and the unconscious, the intensity of my fieldwork on collaborationism became both
rewarding and problematic.

By the time I returned to the town five years later, in the summer of
1987—which also happened to be the summer of the trial of Klaus Barbie,
former head of the Gestapo in Lyon, on charges of crimes against humanity—
the cobbler had passed away. His family had become less enthusiastic about
the idea of my doing a book about him and I could not convince his daughter to talk to me about wartime events. I was recognizing the gap in records
about her father's life, and she was responding: she asked that I omit from my
written account any discussion of the two-and-a-half years of her father's imprisonment, the greatest stigma for the family reputation. After all, what was

such a short period of time out of a life that had spanned 90 years? One day that summer, as we sat together talking about her father, my eye glanced down at the regional newspaper, *La Dépêche du Midi*, which was laid out on the table next to us. At the top was an article on the ongoing Barbie trial. When I asked the cobbler's daughter what she was reading, she answered, "Oh, I'm doing the daily crossword puzzle and checking the lottery numbers." Like many other French people of her generation, she thought the Barbie trial was unnecessarily digging up a past that she had tried hard to forget. The wartime years had seen first the illness of her younger daughter, who became a life-long invalid, and, several years later, her divorce. Following her divorce the cobbler's daughter left her life in Paris and returned to her rural hometown and the family business. After her father's legal judgment, the majority of the townspeople exhibited hostility toward the family, and many boycotted the shoe shop. Very slowly some came around and returned as customers, largely as a result of the daughter's and her mother's careful personal diplomacy.

Toward the end of that summer of 1987, I finally gained the daughter's consent to access her father's poems, jumbled together in several satchels near his cobbler's workbench. Then, in the course of my reading, I made a dramatic discovery. I recognized the concluding four lines from Dupont's autobiographical poem about his childhood spent working on farms, "*Mon temps passat*," which I had discussed extensively with the cobbler six years earlier. In slightly different form, Dupont's concluding lines were the beginning to another poem, also written in Gascon, but penned by a different author, Jean Nadal. I found a handwritten copy of Nadal's poem "*Enrégado o Escoubassot*"[4] sandwiched in between the various drafts and poems in Dupont's satchels. Jean Nadal, a local peasant and Gascon poet, had been an active member of the Militia during the war. He was arrested and tried for collaboration by a military court in the fall of 1944, as was the case throughout France for those persons considered to be the most active collaborators. Judged guilty, Nadal was sentenced to death by firing squad. He had composed "*Enrégado o Escoubassot*" in the weeks prior to his death in December 1944.

During my research in 1981 and 1982, Dupont had spoken to me about Nadal, saying that he had been a great friend of his, a master of the regional language in his verse-making, and that he had been needlessly shot. I, however, avoided discussing Jean Nadal further with Dupont at the time because I was uneasy with Nadal's negative social press as an executed collaborator. Several other town residents had spoken to me about Nadal that same year, echoing Dupont's judgment about Nadal's outstanding linguistic competence in Gascon. As these elderly residents talked with me, they swore me to secrecy about their knowing Nadal, his poems, and how he had died. They asked me not to discuss Nadal further with anyone.

During the summer of 1987, while continuing my linguistic fieldwork, I came up against multiple strands of historical consciousness: What seemed to be a simple retrieval of information on my part, documenting personal history through poems written in Gascon, was in fact imbricating ethnography with national history. Moving within a network of local writers and readers of dialect literature while recirculating Nadal and Dupont's poems, I, the fieldworker, was unwittingly circulating sympathy for the political positions of these two Militia men, at least from Dupont's perspective. Language was a tool, a channel, a method, at the very core of this process of historicizing, for both the cobbler and me.

My discovery of the reinscription of Jean Nadal's poem into Dupont's childhood lifestory poem underscored the cobbler's obsessive reliving of wartime events and provided me with a form of evidence about his past political allegiances. This was a turning point in my evolving understanding of his writings. I spent subsequent years of research looking for other kinds of written evidence in the public domain that would cast light on and confirm Dupont's wartime right-wing political views, such as an article written by Dupont, either in the general regional press or in the more specialized wartime Legionnaire or Militia publications, but I could find no such trace of evidence. Such publications, if available, exist in incomplete series in Paris, in libraries such as the *Bibliothèque Nationale* or the *Institut de l'Histoire du Temps présent*. The cobbler had neither access to the social networks nor the degree of literacy required to publish in such sources. Through the oral histories and Dupont's poems I had gained access to his secret, which as an ethnographer I could obtain, unlike the historian of more distant time periods who cannot speak with his or her sources. It became crucial to me that I use the ethnographic encounter to set the historical record straight.

In November 1996, I consulted the post-war trial records of Jacques Dupont in a departmental archive. The dossier is fascinating and a vivid portrait of the community and the period emerges. The context of the writing, namely a judicial inquiry, gives it a particular cast, of course, and most of the documents it contains fit the genre—they are filed to establish the positive social reputation of the cobbler. These documents included copies of the written testimonies that Dupont had first shown me some 15 years earlier, which he and his lawyer had gathered to counter the collaborationist charges. Predictably, most of the witnesses in defense of Dupont came from the original community, knew him well, and spoke of him favorably. A number of people sympathetic to and/or active in the local Resistance testified in his behalf. The cobbler's wartime involvement, first with the Legion and SOL, then with the Militia, was presented as a lapse in good judgment, with many witnesses expressing confusion between their own public image of the cobbler as a solid and obliging local citizen and his privately held motivations for joining the Militia.

Certainly, major ideological shifts took place between the French Legion, founded in 1940, and the Militia of January 1943. The former group actively recruited World War I veterans throughout rural France—men who had been too old to serve in the brief Franco-German conflict of 1939 that had ended in France's defeat. Stung by this military humiliation, the veterans still held a cult for Pétain as their commanding officer from World War I. The Legion promoted a broad patriotism and military action as a solution to political conflict, but was not directly collaborationist at its inception. As the ideological climate intensified, so did the recruitment of members into other groups such as the SOL, a more hard-core version of the Legion, and finally into the Militia, the most openly collaborationist and violent of the three. The Militia's elite unit trained with Nazi officers from the Waffen-SS, one of the two major divisions of the German Armed Forces, before departing for combat on the Russian front. Dupont had always insisted to me in our one-on-one conversations that by the time he realized the true nature of the Militia's activities, it was too late to quit. He stayed on in the Militia, he insisted, to serve as a cover for his son-in-law's underground Resistance activities, a point that he made in his trial depositions. The details of such Resistance activities were never mentioned, however, and I found no account of them in Dupont's papers.

The role of Dupont's son-in-law in local events is far from clear. A military officer from an upper middle-class family in Paris, he came to the town in 1940 to visit a male buddy; here he met and then married the cobbler's daughter the following year. He was a key figure in recruiting young people for the Legion. A tireless propagandist, he published several articles in one of the Legion's regional journals, *Reconstruire,* in late 1943 and early 1944. The articles are polemical historical pieces and indirectly frame events of the time via the mechanism of colonization and France's historical civilizing mission. Dupont's son-in-law appears to have changed sides from Legion member to Resistor in early 1944, and he participated in the liberation of Paris that fall. He then went off to Indochina, as did many collaborators, to defend French colonial interests. Questions remain. Did he truly believe in the principles of the Legion and what did he know about the deportations? What made him turn from the Legion to the Resistance (which Resistance?) and to what extent did this involve his father-in-law the cobbler? Was he an opportunist who sensed that when defeat was imminent he would need the life insurance of being labeled a Resistor (a *Résistant de la onzième heure*)? Was he simply a career military man, always ready to defend "Frenchness," no matter where and how?

Dupont's use of the Militia as a political cover seems unlikely, even if it perpetuates a family myth of male bonding. More realistic is that Dupont was used by his son-in-law, who left the family and the region after the war. Motivations for joining the Militia were varied, and, in small local communities social networks based on kinship, neighborhood, and patron-client affiliations were often a key factor in constituting Militia membership. Until recently, relatively few studies of the Militia have appeared in print (Gordon 1980; Delperrie de Bayac 1994; Germain 1997; Giolitto 1997; Terrisse 1997). It took me a long time to face what Dupont's membership in the Militia could mean, for it represented the worst part of the war—the aspect that I wanted least to confront. It was only as I read historians' studies about the Militia throughout the 1990s that I understood what levels of intimidation and violence its members had practiced against civilians suspected of supporting efforts of the Resistance.

CRIMES AND PUNISHMENTS: ASSESSING GUILT

At his post-war trial, Dupont was accused of three crimes. First, he was accused of having borne arms against France, stemming from his guarding the town post office the night after the Allied landing with a gun provided by his Militia superiors, who had barricaded the town against the arrivals of the local Resistors as they emerged from their hiding places in the underground. Witnesses testified that they overheard Dupont object to this order and attempt to refuse to follow it. In the end, the jury agreed that this sole instance of a forced bearing of arms was not symptomatic of a larger military action, and they judged Dupont not guilty of this charge. Second, the cobbler was accused of having "denounced" two young people who had mocked the image of Marshall Pétain in a news clip shown at the town movie theater. In Vichy France, commercial films, which were selected to reflect the larger ideological values of the régime—of "*Travail, Famille, Patrie*" ("Work, Family, and Fatherland")—typically contained an official propaganda piece at the end of their showing. The names of the two young men were then communicated to the local Militia, who in turn gave their names to the *Service du Travail Obligatoire*. This organization, the STO, conscripted French citizens to work in Germany as part of the Nazi war effort in exchange for the liberation of French prisoners of war. The names of the two young men never appear in the trial record, which also does not indicate what happened to them. Were they indeed sent to the STO, or did they go underground to join the Resistance, a common response at the time? On this second charge Dupont was judged guilty by a majority vote of the jury.

Third, Dupont was accused of having informed his Militia superiors of at least one Allied parachute landing of equipment that had been destined for the local Resistance, and thus rerouting arms and supplies needed for the French anti-Nazi effort into the wrong hands. Dupont was also judged guilty on this charge by a majority vote of the jury. The details surrounding the parachute landing in question still appear very murky, in light of what I have been able to piece together. What further complicates matters is that the cobbler served both sides by acting as a double agent. Although ideologically sympathetic to the principles of the Legion and of the Militia, he enjoyed a strong personal friendship with the key figure of the local Resistance, Alain Garnier, who was executed by the Germans in a village near the Pyrenees in July 1944. No one was ever able to specify the extent of Dupont's aid to local Resistors and to Garnier, including Garnier's widow, who testified on the cobbler's behalf at his trial. The sole survivor of the German attack on Garnier's unit, fellow towns-man Bernard Petit, whom I interviewed in the spring of 1997, could not say much about Dupont and Garnier's relationship either, beyond his intuition that each had a deep loyalty to look out for the other. Petit underscored the guerrilla warfare tactics of his Resistance unit, which included constant moves, the use of first names or pseudonyms to cloak civilian identities, and a degree of secrecy that resulted in no one knowing exactly what his or her neighbor—in the Resistance, in whatever group—was up to.

There are other means by which to retroactively evaluate any active po-litical collaboration by Dupont, which would involve more research, espe-cially in reconstituting the local actions of the Militia during the War. This localized element is difficult to determine, since many written documents were destroyed by the *Miliciens* as the conflict came to a close (Gordon 1980:361). One important element of Militia actions involved arresting and assisting in the deportations of Jews, both French and foreign-born. Dupont's hometown, like many others in southern France, was the temporary residence of Jewish families and individuals in transit to Spain, three hours to the south. While the majority of the Jews residing in the town escaped deportation, at least four men were arrested and deported between 1942 and 1944. There is no public record of this in the wartime regional press, although there are other, increasingly sinister descriptions of charges brought against foreign-born Jews, Spaniards, and Italians during this same time period.[5]

In the course of his trial, Dupont insisted that he did not play an active role in the Militia, including making denunciations and arrests; instead, he merely collected membership fees. Other private correspondence of his, which I have consulted separately from the trial records, suggests that he supported the Militia, at least ideologically. It also appears that two higher-ranking Militia officials, the previously-mentioned poet Jean Nadal and another man, Gilbert Pinel, headed up the political action taken by the local unit. Dupont was friendly with both men; the former was a peasant and the latter a baker, two professions that paralleled Dupont's working-class identity of cobbler. After being brought to trial by the military courts in the fall of 1944, first Gilbert Pinel and then Jean Nadal were executed within several months of each other.

The charge of collaborator brought against Dupont may be understood at a more symbolic level as the bad ending to a fairy tale. The cobbler's daughter had married up, well out of her social class. Going after her father may have served as a local lesson for others who tried to similarly challenge the social hierarchy, as one town resident observed to me. In fact, Dupont was one of the few, and perhaps the only, town resident(s) judged for collaborationism to have actually served his prison term, which was reduced from five to two-and-a-half years for "good behavior." He spent many years working to pay off the 10,000-franc fine assessed at the time of his sentence. In contrast, many individuals accused of collaborationism and brought to trial did not actually serve out their sentences. Those in higher positions on the social ladder, who engaged bigger-time city lawyers, managed to obtain lesser sentences. Others fled, either to a neighboring department—where some lived openly while others remained within the confines of convent or monastery walls[6]—or even further to other countries, such as Germany or Chile. South America would prove to be an effective hiding place for some of the most prominent collaborators, including Klaus Barbie, head of the Lyon Gestapo. Historian Robert Paxton notes that retribution for male collaborators differed according to groups, such that experts, businessmen, and bureaucrats survived almost intact (Gordon 1996: 183), while the purge struck much harder at men of words (journalists, commentators, intellectuals, artists) whose published documents were used by courts to judge them (Golsan 2000:182).

NEGOTIATING PUBLIC AND PRIVATE IDENTITIES

La vie privée est, dans tous les cas, un rempart que nul ne peut passer. Pas même l'historien.

—Jean-Denis Bredin (1984: 97)

"Private life is, in all cases, a barrier that no one can overcome. Not even the historian," says Jean-Denis Bredin, a scholar of the Dreyfus Affair, reflecting on the relationship between the law, the judge, and the historian. I might add "and not even the ethnographer" in talking about the French context and Jacques Dupont. French state restrictions insist on a waiting period of 30 to 150 years for consulting public archives, depending on the category of archives and the importance that the law ascribes to the protected secrets contained within them.[7] Prior to the fall of 1997, scholars interested in the World War II period had to go through the special appeals process of filing an exception in order to access archival materials. This process was significantly changed with the advent of Socialist Lionel Jospin as France's prime minister in May 1997. Article 6.6 of the October 2, 1997, Jospin circular, which outlines access to public archives for the 1940–1945 period, reads: "This is why, without any exception, of which I would like my office to be informed, requests for access to archives having to do with the period 1940–1945 should no longer be denied on the basis of these grounds [namely, previously cited issues of the security of the state and of national defense]. Consequently, the only rationale on which a refusal for an exception can be based is respect for the private domain. But, even in this case, refusals must not be systematic. In particular, members of the scientific community whose serious and honorable qualities are recognized should be able to have access to these documents if they swear in writing to preserve the anonymity of the persons in question and to make no non-historical use of information they have discovered" (*Journal officiel,* October 3, 1997: 14340; my translation). "The anonymity of the person" refers to protecting individuals' social reputations, including posthumously, by changing their family names. "Non-historical use of information," broadly conceived, refers to those texts about persons judged during the Occupation composed by scholars in the historical present and circulated beyond academia to the broader media, including the print press.

I have focused on the details stemming from one case of collaborationism so as to provide insight into the complexities of political life in rural France during World War II, its historical reconstruction, and degree of access to wartime archives and the French legal system. The issue of representativeness arises. Dupont shares a number of social characteristics of those individuals of his time who were drawn to fascism. His social class, that of small shopkeepers and craftspeople, saw the nature of work change with the advent of modernization and mass production and, in the throes of economic crisis, they identified with the fascist call to return to an earlier, preindustrial world. The cobbler's generation, that of World War One veterans, had witnessed first-hand the definitive separation of Church and State in 1905, an issue that had polarized political forces on the French Left and

Right in the late nineteenth century. This same generation had been schooled in a heightened French patriotism dating back to its elementary school lessons, which called for revenge after the French defeat in the brief Franco-Prussian conflict of 1870. Dupont's identification with the spirit of revenge, part of his motivation to sign up to fight during World War I while still a legal minor, would take a different form during World War II. Too old to fight, he sought political involvement through membership in organizations whose increasingly fascist profile could serve as a new locus of ideology and values.

Dupont's political discourse, that of a person with a limited grade-school education, was not always nuanced. Reflecting the propaganda writing of the time, Dupont would speak to me of the "terrorists" of World War II, a category in which he lumped together Resistors, communists, and Gaullists, and from which he clearly distanced himself. At the same time, he consciously manipulated two other identity markers. First, he frequently referred to himself as a Spaniard, perhaps to underscore being an outsider, rather than a Frenchman who had betrayed his country. Second, he assured me that he had been a *Pétainiste*—simply a loyal follower of Pétain, the former World War I commander of French military forces—and not a *collaborateur*. The latter distinction was operative in the internment camp where Dupont resided for several months in the fall of 1945 before serving his prison sentence in a nearby city. It parallels Gordon's 1980 distinction between collaborationists and collaborators, the former term emphasizing ideology and the latter direct political action.

While a neighbor asserted that during the war Dupont "did not like Jews," this attitude appeared to crystallize in a stereotypical economic anti-Semitism that was shared by many different social classes in France, within and beyond the region. Dupont's general ignorance of Jewish culture was not atypical for the rural southwest, which had few or no Jewish families. Historically, areas with significant Jewish populations include Avignon, in Provence; Bayonne, a city south of Bordeaux, bordering the French Basque country; and the northeastern region of Alsace. Local people frequently commented to me in the course of my interviews that until the war they did not know what a Jew was. Dupont's public and private remarks, oral and written, did not take on the virulently racist overtones of a prominent collaborator such as Paul Touvier, former head of the Militia for the Rhône-Alps region. Touvier's unpublished diary, presented at his trial, was a key piece of evidence that condemned its author as maintaining an unchanged political position over time, one of deepest anti-Semitism, despite Touvier's verbal protestations to the contrary (Golsan 1996:41).

It would be tempting to be able to definitively tie up the loose ends in the cobbler's case and decide once and for all his degree of complicity in wartime events. I have sought to do this over these many years and so finish a dialogue that I never completed with Dupont. The complexity of interpreting wartime events, both real and remembered, and of labeling Collaborators and Resistors, resurfaced for me in a different political context, during my 1997 interview with Resistor Bernard Petit. Born in the same town as Dupont, a fellow town resident, and a retired furniture-maker, Petit wept as he recounted his escape from his unit's hiding place on a farm on the outskirts of a Pyrenean village in July 1944. He explained that some of his comrades, who were from more middle-class backgrounds than his own rough and tumble working-class life, had opted to sleep in the main house on real beds, and so were unable to escape as the Germans drew in on them. He and two others had chosen the barn, however, contenting themselves with makeshift beds of straw pallets. When their small unit of Resistors heard the Germans surround the house, Bernard Petit could think only of his survival. He tore out of the barn, crossed the yard, and jumped over the fence in the midst of a volley of shots and, in a superhuman effort, made it past the armed Nazis into the woods. Alain Garnier and the three other men died.

At the end of her husband's narration, Mrs. Petit said to me with bitterness that townspeople would have preferred to see him die, gunned down as a martyr for the cause, and that their family had borne the brunt of survivor's guilt ever since. Local rumor even had it that her husband had denounced his fellow Resistors and given away the location of their hideout, which resulted in Petit being shunned by much of the community for the rest of his life. Petit's moral condemnation, based largely on false rumor, shows how such a mechanism can work regardless of ideological orientation, and to what extent opinions could vary (historically) as to what constituted being a "Resistor."

Let me return to the question of sources and the complex place of the ethnographer, whose role it is to step back from the minutiae of everyday life, contextualize, and do cross-checks of different forms of evidence, ranging from oral accounts to court records and newspaper articles. Throughout my fieldwork year Dupont obsessively spoke to me about the war, a point that I found confirmed both on my audiotapes, which I listened to years later, and in analyzing many of his writings. I read those writings—which the cobbler had shared openly with me during the 1981–1982 fieldwork year—over and over again. After his death, however, I could not convince his family to talk about them with me.

Unable to return to the critical war testimonies that Dupont had first shown me, I went through a different, official channel—that of academic scholar—to gain access to copies of those records. But this action, while bringing part of a chapter of my work to a close, opens up new paths. His daughter's role appears ambiguous. Was she just a military wife concentrating on raising her two young daughters? How actively did she participate in the events of the day? Is that another reason why she will not talk with me about those wartime years? Understandably, in turn, her elder daughter and family do not want their reputation sullied. There is more pain here. The cobbler's granddaughter had little contact with her father after her adolescent years, and for most of her life it was her maternal grandfather, the cobbler, who took on the role of father figure. Born during the war, her generation grew up with the dramatic consequences of family members' actions, much of which was kept silenced, as was the case for the children and grandchildren of Nazi Germany.

The ethnographer is subject as well to taboos beyond those of the cobbler's immediate family. Much of the community turns away from those difficult days, for its inhabitants all had to live with one another and remake some kind of life for themselves after the war. Born after World War II and from a different country, I am not part of the local world. It had taken me quite a while to locate Mr. Matet, the departmental correspondent for the *Comité d'Histoire de la Seconde Guerre mondiale* who lives in a nearby town; when I did, he did not want to give me access to his files on Resistors or Collaborators, painstakingly put together through years of interviews. I may be an insider ideologically, sharing his revulsion for collaborationism, but, as an American, I am an outsider. Petit, the surviving Resistor, observed wryly that local townspeople might be amused and "finally get their butts in gear to write their own history," as he put it, if they see me, an American, beating them to it. I am positioned in the community, whether I like it or not. Beyond the local community, ethnography becomes a factor in shaping if/how/and why a history ever gets told.

AMNESTY AND ABSOLUTION

What were Dupont's actions and how should he be judged for them? The charges brought against the cobbler seem weighted since others were freed on far more damaging evidence. And what do the charges represent: trumped-up statements by personal enemies—in which case Dupont was innocent but found guilty—or, instead, even worse crimes which do not make it into this particular written record—for which he was judged guilty but not found sufficiently guilty? The charges and the punishment suggest the arbitrariness of justice, especially when it involves working-class people.

Why should we assume that the cobbler's private writings are definitive proof of his political positions, and that there is a correlation between autobiographical writing and truth? How does degree of intentionality figure in? Gaps between the written and the oral, the past and the present, and words and actions make it difficult to determine Dupont's guilt. Furthermore, over time, history often elides the distinctions between hero and traitor in a subtle process of revisionism that installs a coherent, determinist and reductionist discourse through a two-stage process, first by questioning the political principles on which decisions were made at the time—"after all, all the Resistors were bums" or "where does collaboration begin and end"—and then by relativizing, rendering banal and negating the evil actions of the past: "evil was on both sides" (Benasayagin 1994:250–251).

For the time being, it turns out that the group of the three monkeys has the last word. The French have a specific guarantee for protecting national consensus, namely the amnesty, whereby political crimes and their punishments are deemed no longer to exist (Gacon 1994:100–102). The French press law of 1881, still in force, makes it illegal to publish or discuss in public any punishments, including those that have been amnestied. That the published acts are true is no defense in cases involving amnestied crimes, or, for that matter, any legal punishments that are more than ten years old. This statute considerably constrains the ability of the historian to present factual evidence, notes Bredin in his trenchant critique of the French legal system (1984:109). The French press law has a chilling effect on editors, making them very cautious about printing historic facts that could involve them in costly, prolonged lawsuits, even if they are likely to win. Moreover, the ban does not end with the death of the amnestied person, for the family may sue for harm to "the honor or the reputation of the living heirs or spouses." In practice, no one is sued for discussing major historical figures such as Philippe Pétain, Pierre Laval, or Joseph Darnand, the chief of Vichy's Militia. But recalling the past or the post-war punishment of a simple Militiaman could lead to a lawsuit (Koven 1995:349).

In November 1997, after having been granted another special permission to consult contemporary historical archives, I examined the cobbler's application for a presidential pardon and amnesty, the second part of his legal dossier. At the Archives of the Ministry of Justice in Paris, I reencountered many of the documents from the post-war trial. I found additional papers that attested to Dupont's honorable behavior and his repeated use of a family defense strategy—"I need to support my family" and "I need to restore my good family name"—to argue for early release from prison and legal rehabilitation. The amnesty was finally granted in May 1951, as the last round of individuals judged for collaborationism after the war received general pardons from President Vincent Auriol. It turns out that Dupont had also insisted on, and finally obtained, an individual amnesty. From a legal point of view, this means that his entire political past was erased. He could again function as a full-fledged citizen in French society, and his right to vote was restored. But Dupont could not shake himself free from the consequences of the wartime judgment. Consequently, much of his verse-making in the final two decades of his life was devoted to rewriting his self-image for himself and family members as well as for selected others within his hometown, even as collective memory faded about his exact role in the town's past.

In assessing the social judgment of collaborators at the local level, one functionalist view considers them as exemplifying a breakdown in village social structure, which demands solidarity: Within French village society, one must do what is done and not do what is not done. Those who do not respect this tacit rule run the risk of criticism, even a type of exclusion: Not playing the game puts a person out of the game.[8] Yet, this village solidarity is not seamless. The secret, a condition for the very existence of interpersonal relations, is part of the social structure and cannot always be revealed.[9] It transcends those categories so well known to the ethnographer, of gender, race, caste, and religious affiliation, for example. As such, the secret is an integral part of the local linguistic economy, moving between speech and silence and situating individuals across time.

One obvious conclusion to Dupont's story is that the power structure reproduces itself: the most guilty and powerful maneuver to discharge their guilt, while the less guilty and poor live with and in the legal system's judgment. The cobbler as simple working man was not so simple, however, nor were his legal strategies. He requested, and was granted, a broad pardon; he requested, and was granted, an individual pardon. The individuals linked to Dupont's case—Dupont, his family, the local community, and the ethnographer—also remind one of the multiple meanings of historical events. The legal judgment defined Dupont for almost half of his life, a good 40 years, and this well after the events in question. He and I have both been obsessed by this judgment, for quite different reasons: he to erase this mark on his social reputation, I to assess his actions during the war and use ethnographic practice as a tool to negotiate the past and its telling.

In reading through Dupont's words, including those by and about him contained within the legal dossiers, I am reminded that legal accounts are in part stories of events, claimed or ratified to be "true" but bearing fictional element(s).[10] In his position as a small-town cobbler and folk poet, Dupont was excluded from local and national elite literary circles that privilege knowledge of French, the national language. He could both talk through and hide in his Gascon poems, as could fellow Gascon poet Jean Nadal. Folk poetry is hidden, and folk poetry as political evidence is hidden, with implications for evidential understanding in legal cases. Throughout my 1981–1982 research year, the cobbler understood this point, when I did not yet. Thus, toward the end of my field stay, before I left the town to return to the States, Dupont joked with me, "What will you do with my story, denounce me?" After his death, his family was uncomfortable: "It seems that my father talked to you about all kinds of things that we never discussed" said his daughter, and, "Do you only see my father in terms of the war" during my 1987 summer field-work. Her comments underscore the biographer's bind: family members live their own lives often close to their communities of origin, into and from which the ethnographer steps by choice. Her comments also underscore a certain disciplinary reality: Much of French historical scholarship about the war has tended to focus on groups, not individuals. Ethnography, in contrast, offers a method and an approach: It permits fine-grained, close readings of social situations, individual lives, and definitional crises.

EPILOGUE:
A RECENT FRENCH WORLD WAR TWO TRIAL

I have taken my readers through a series of stories about a working-class man in order to show the interpenetration of oral and written sources, the role of multiple witness accounts, the power of representation and the class-related arbitrariness of legal justice. By way of contrast, and conclusion, I evoke the 1997 trial of a political big man, Maurice Papon, on the charges of crimes against humanity, and its April 2002 update. Papon was second in command to the regional prefect Maurice Sabatier; as secretary general for the prefecture of the Gironde department, in the city of Bordeaux in southwestern France, during the war he signed thousands of official documents, including orders to deport Jews. Ten different convoys took place over a period of two years—between July 18, 1942 and May 13, 1944— which sent some 1,500 Jews, including more than 200 children, from the detention camp at Mérignac, on the outskirts of Bordeaux, to the transit camp at Drancy, just outside of Paris, then to their deaths at Auschwitz, which was specified on deportation orders as "*destination inconnue*" (destination unknown). While Papon acknowledged post-facto at his trial the unjust persecution of the Jews during the war, he insisted that, at the time, in his role as secretary general of the prefecture, he was not aware of the true nature of the death camps and that he was not alone in this ignorance. His claim has been disputed by a number of World War II historians, including Robert Paxton.[11]

Papon was the first French government official brought to trial on the charges of crimes against humanity;[12] the central issue was whether his bureaucratic activity could be judged as criminal. In contrast to the previous trials of two individuals judged for their role during the war—that of Klaus Barbie, former head of the Gestapo in Lyon, in 1987 and that of Paul Touvier, former head of the Militia in Lyon, in 1994—the case of Papon was even more dramatic, as it involved the French state. After the war, Papon continued to play a prominent political role regionally in France, then in the French North African colonies, and finally nationally, most notably as head of the Prefecture of Police in Paris for President Charles de Gaulle, from 1958 through 1966, and as Budget Minister for President Valéry Giscard d'Estaing, from 1978 through 1981. Thus, to put on trial the supremely efficient civil servant Papon, was a means by which to show that "the apparatus of the State was responsible" for its actions during the war (Gopnik 1998:87; Golsan 2000).

Nuances of degree of moral involvement, complicity, and accountability were at the heart of debates heard during the trial, presented by both the defendant and the prosecution, and would figure into how Papon was judged. After a lengthy six-month trial, which saw considerable wrangling about what the terms "Collaborator" and "Resistor" meant, on April 2, 1998, Maurice Papon was judged partially guilty of complicity in the Bordeaux deportations. Papon's sentence, similar to those passed first by military and then by civil courts during the immediate post-war purges, included ten years of imprisonment, being stripped of his civil and family rights, and being assessed a significant monetary fine. One disappointed journalist observed that a ten-year prison term is what a breaking and entering conviction in France brings (Gopnik 1998:95). Papon and his lawyers immediately appealed his legal judgment, and the defendant attempted to avoid serving his sentence by hiding out in Switzerland. Returned to France, he began to serve his prison sentence on October 22, 1999. On September 4, 2001, Maurice Papon was brought up on charges of illegally arranging for personal bankruptcy by passing his estate on to his children, thus avoiding payment of his assessed fine (Simonnot 2002). To date, Papon has not paid a cent to the civil parties of the 1997 lawsuit.

Papon next requested through his lawyers, and was denied, a presidential pardon (on October 9, 2001). His lawyers then brought his case before the European Court of the Rights of Man, where they successfully obtained a judgment of state guilt: Maurice Papon, individual French citizen, could not be judged independently of the collective body of the French state—both would share equal accountability for war-time actions. On April 12, 2002, the French Council of State ruled that the French state would now pay one half of the fine assessed to Papon at the 1998 ruling.[13] The 2002 ruling is a reminder of how responsibility for individual actions can be strategically shifted back onto the nation-state. Papon declared himself quite satisfied with the outcome, noting that he should not have been singled out to serve as a scapegoat for the nation. Overlapping legal jurisdictions aided him, for what allowed Papon to overturn part of the French court's judgment was an appeal to legal subtleties within a different, supranational jurisdiction—that of the European Court of the Rights of Man.[14]

The immediate post-war trials of collaborators, and the much later trials of accessories to crimes against humanity, have a critical memory function, as they attempt to restore the individual faces of the victims of the Holocaust. Yet these trials contain a painful paradox: By making the French state accountable and thereby establishing a continuity between the historical past and the political/judicial present, the degree of individual guilt, such as that of the bureaucrat Papon, is lessened. I close with several questions: How do statutes of limitation for crimes serve citizens and the collective social body? How does society constitute itself after civil war? What is the weight of the individual's actions? When is the individual seen as part of the collective body and thus not held accountable? On this last point, the amnesty articles of the French 1951 penal code provided many loopholes to persons judged as collaborators.

NOTES

1. This name, like all others in this chapter that pertain to individuals who are not in the public domain, has been changed. I would like to acknowledge the many people who have helped me in my research in France since 1981. They include directors and staff members of regional and departmental archives in southwestern France, in national archives and research libraries in Paris, university academics, and men and women from a number of communities, rural and urban, in the southwest. The original 1981–1982 research was funded by a Wenner-Gren grant; the 1987 research was funded by a grant from the Institute of Arts and Humanistic Studies at Penn State University. Stephanie Kane and Phil Parnell provided critical editorial feedback.

2. The title of the cobbler's poem translates doubly, as "My past times" or "My time spent."

3. Exemplified by the emphasis on the majority of the French people as Resistors, a myth actively propagated by Charles de Gaulle in the post-war years, which

discounted the ambiguities of individuals' political positions, and downplayed the Collaboration. Restrictions on access to wartime archives for researchers would also contribute to the air of secrecy surrounding this time period.

4. This title translates literally as "End or Beginning" and references Nadal's discussion of his impending death in Christian terms.

5. In a departmental version of the Toulouse-based *La Dépêche,* dated Thursday, March 12, 1942, appears the following: "Black market. Searches have made recently in the homes of certain Jews living in M. Large stocks of merchandise were discovered. Their owners will be pursed legally on the grounds of non-declaration." On Tuesday, April 28, 1942: "Nina P. and Anna B. (Polish Jewish family names) living in M. without a visa or identity card, are fined 300 francs each and the first person an additional 50 francs for lack of a pass." On Thursday, June 8, 1944, two days after the Allied landing: "Correctional Tribunal. For not renewing a foreign identity card, Susana T., married name B., 300 francs fine, with delayed payment, Caterina S. (both Italian family names), 300 francs fine, with delayed payment."

6. This was the case for Paul Touvier, whose flight from justice lasted more than 40 years and would have been impossible without the help of Catholic clerics and powerful groups within the Church (Golsan 1996:28–29; 34–35). During an interview in spring 1997, one of the former residents from a village close to Mr. Dupont's hometown told me how she and her husband had run across Jean Nadal's widow quite by accident, 20 years after the war, in a Bordeaux open-air market. It turned out that Nadal's widow was living in a convent under an assumed name, that of her mother's first name. She spent the remaining years of her life working as a cook for the convent.

7. In his chapter "Secrets de l'histoire et histoire du secret," Gérard Vincent quotes from the law of January 3, 1979, and the decree of December 3, 1979, which regulate the minimum 30-year waiting period for public archives to 60, 100, or even 150 years, depending on the importance that the law ascribes to the protected secret. While the archival administration can shorten this waiting period, the unchangeable conditions remain the 100-year waiting period that protects "individual information having to do with personal and family life and facts and behavior of a private nature" and the 150-year waiting period when the information is "medical in nature" (1985: 163). For the French version of article seven of the January 3, 1979, law that describes these various waiting periods, see note two in Bredin (1984).

8. My translation of historian Antoine Prost's discussion of the transition between private and public life (1985:116).

9. For a remarkable synthesis of a range of sources—literary, historical, philosophical, legal—on the general problem of secrecy, primarily in the American context, see Bok (1989).

10. Natalie Zemon Davis (1987) details the relationship between the law, the event, fiction, and truth in the introduction to her historical study of a related genre, sixteenth-century letters of remission, pardon tales that were addressed to the king by peasants and artisans, asking for absolution from their crimes.

11. See citations of Maurice Papon in the articles in *Le Monde* dated October 18, 1997; November 6, 1997; November 16–17, 1997; and December 10, 1997. See "Those Who Organized The Trains Knew There Would Be Deaths," an interview with Robert O. Paxton from the October 3, 1997, issue of the French newspaper *Libération,* reprinted in Golsan (2000:179–183).

12. As defined in the Charter of the International Military Tribunal signed August 8, 1945 in London, crimes against humanity consisted of "murder, extermination, enslavement, deportation, and other inhumane acts committed against any civilian population, before or during the war, or persecutions on political, racial or religious grounds." Those considered liable under law included "leaders, organizers, instigators and accomplices participating in the formulation or execution of a common plan or conspiracy to commit any of the foregoing crimes." Forty years later, in December 1985, the French court of appeals determined that henceforth crimes against humanity would include "inhuman acts and persecutions committed in the name of a State practicing a politics of ideological hegemony [which] were carried out in a systematic fashion not only against persons by reason of their appurtenance to a racial or religious collectivity but also against adversaries of this politics regardless of the form of this opposition." Apart from expanding the definition of those to be included as victims of crimes against humanity, the new legislation also offered a more specific codification of the type of regime in whose name crimes against humanity could be carried out—a state "practicing a politics of ideological hegemony" (see Golsan's edited volume on the Bousquet and Touvier cases [1996:18–19]). In her reconstruction of the legal implications of the Touvier and Papon trials, Leila Sadat observes the efficacy of the Nuremberg international tribunal, whereas the application of international law to new defendants decades after the war had ended by France as a matter of French *municipal* law strained the legal construction of the crime against humanity nearly to the breaking point (2000:146). Sadat concludes that crime against humanity was originally the creation of the international community, and its ultimate force lies in future application by the international community, and, one day, by a permanent criminal court (2000:148).

13. The original assessment was for one-third of the fine (200,000 Euros).

14. See all of Golsan's 2000 edited volume on the Papon case. In particular, see Leila Nadya Sadat's chapter "The Legal Legacy of Maurice Papon" (pp.131–160).

BIBLIOGRAPHY

Autrement
1994 Oublier nos crimes: L'amnésie nationale, une spécificité française? Dimitri Nicolaïdis, ed. Mutations 144 (April).

Benasayagin, Miguel
1994 A qui profite le crime? Autrement 144: 244–254.

Bok, Sissela
1989 (1983)   Secrets: On the Ethics of Concealment and Revelation. New York: Vintage Press.

Bredin, Jean-Denis
1984 Le Droit, Le Juge et L'Historien. Le Débat 32: 93–111.

Code Pénal (Amnistie)
1951 Law 51–18, January 5; 1953, Law 53–681, August 6; 1959, Ordinance 59–199, January 31.

Davis, Natalie Z.
1987 Fiction in the Archives: Pardon Tales and Their Tellers in Sixteenth-Century France. Stanford, CA: Stanford University Press.

Delperrie de Bayac, Jacques
1994 (1969)   Histoire de la Milice, 1918–1945. Paris: Fayard.

Gacon, Stéphane
1994 L'oubli institutionnel. Autrement 144: 98–111.

Germain, Michel
1997 Histoire de la milice et les forces du maintien de l'ordre: guerre civile en Haute-Savoie. Montmélian: La Fontaine de Siloe.

Giolitto, Pierre
1997 Histoire de la milice. Saint Amand-Montrond: Imprimerie BCI.

Golsan, Richard J., ed.
1996 Memory, the Holocaust, and French Justice: The Bousquet and Touvier Affairs. Hanover, NH: University Press of New England.
2000 The Papon Affair: Memory and Justice on Trial. New York: Routledge.

Gopnik, Adam
1998 Papon's Paper Trial. The New Yorker, April 27-May 4: 86–95.

Gordon, Bertram M.
1980 Collaborationism in France during the Second World War. Ithaca, NY: Cornell University Press.
1996 Afterword: Who Were the Guilty and Should They Be Tried? In Memory, the Holocaust, and French Justice: The Bousquet and Touvier Affairs. Richard Golsan, ed. Pp. 179–198. Hanover, NH: University Press of New England.

Journal Officiel de la République Française
1977 October 3.

Kaplan, Alice
2000 The Collaborator: The Trial and Execution of Robert Brasillach. Chicago: University of Chicago Press.

Koven, Ronald.
1995   The Duty to Remember, the Need to Forget. *In* Travelers' Tales. James
       O'Reilly, Larry Habegger, and Sean O'Reilly, eds. Pp. 345–355. San Fran-
       cisco: Travelers' Tales, Inc.
Le Monde
1998   Special issue on the Papon trial, March.
Lévy-Willard, Annette and Béatrice Vallaeys
1997   An Interview with Robert O. Paxton. Libération, October 3. Reprinted as
       Those Who Organized The Trains Knew There Would Be Deaths. *In* The
       Papon Affair: Memory and Justice on Trial. Richard J. Golsan, ed. Pp.
       179–183. New York: Routledge.
Prost, Antoine
1985   Transitions et interferences. *In* Histoire de la vie privée: De la Première
       Guerre mondiale à nos jours. Philippe Ariès and Georges Duby, eds. Tome 5.
       Pp. 115–153. Paris: Seuil.
Reconstruire: Circulaire de Liaison de la Jeune Légion Gasconne
December 1943-January 1944.
Sadat, Leila Nadya
2000   The Legal Legacy of Maurice Papon. *In* The Papon Affair: Memory and Jus-
       tice on Trial. Richard J. Golsan, ed. Pp. 131–160. New York: Routledge.
Simonnot, Dominique
2002   L'Etat devra partager la condamnation de Papon; Le Conseil d'Etat reconnaît la
       responsabilité de Vichy in Libération. April 13. P. 17 (http://www.chez.com/
       constit/papon.html).
Taylor, Archer
1996   "Audi, Vide, Tace," and the Three Monkeys. De Proverbio 2 (1): 165–171.
       Tasmania, Australia: University of Tasmania.
Terrisse, René
1997   La milice à Bordeaux. Bordeaux: Auberon.
Vincent, Gérard
1985   Secrets de l'histoire et histoire du secret. *In* Histoire de la vie privée: De la
       Première Guerre mondiale à nos jours, Philippe Ariès and Georges Duby, eds.
       Pp. 158–199. Tome 5. Paris: Seuil.

CHAPTER 11

SOLIDARITY AND OBJECTIVITY
Re-Reading Durkheim

Carol J. Greenhouse

We must not say that an action shocks the common conscience because it is
criminal, but rather that it is criminal because it shocks the common conscience.
—Emile Durkheim, *The Division of Labor in Society*

TO UNDERTAKE AN ETHNOGRAPHY OF CRIME is to probe the relation between
science and ethics at their mutual points of extremity: where law might kill
on the basis of someone's judgments or where private ethical reflection is
most pressed against public claims of universal moral principle.[1] For
Durkheim, the "objective element" (Durkheim 1933:36–37) of sociology—
the point of exactitude from which scientific argument exhorts its readers—
originates in the minds of willing individuals. There, thinkable futures await
their naming, occasioning (or not) a shift in the collective conscience. At
that point of possibility—a possibility that takes the form of a communicative exchange—Durkheim locates the ground for his "sociological method."

Indeed, Durkheim's sociological method assumes both a willing scientist and a willing reader who will always prefer rationality over irrationality,
and compassion over justice; it also assumes a particular form of sociological authority. Sociological authority rests not on observation, Durkheim
writes, but on something else—on the commitment to discerning "the way
in which [facts] are scientific": "To subject an order of facts to science," he

writes, "it is not sufficient to observe them carefully, to describe and classify them, but what is a great deal more difficult, we must also find . . . *the way in which they are scientific,* that is to say, to discover in them some objective element that allows an exact determination, and if possible, measurement" (Durkheim 1933: 36–37; original emphasis).

The reason to undertake an ethnography of crime, then, is not just to perfect our knowledge of crime and punishment, but also to maintain the vital connection between the practice of social science and the question of its fundamental social value. In that spirit, what follows is not an ethnography of crime, but a brief chronicle of reflection and re-reading.

Some years ago, after perestroika but well before the Yeltsin coup, the American Bar Association initiated an exchange program that brought some 30 Soviet lawyers to law schools and law firms in the United States for a year's visit. In this same historical context, Slavoj Žižek asked: "Why is the West so fascinated by recent events in Eastern Europe?" He answered his own question this way:

> The answer seems obvious: what fascinates the Western gaze is the reinvention of democracy. It is as if democracy, which in the West shows increasing signs of decay and crisis, lost in bureaucratic routine and publicity-style election campaigns, is being rediscovered in Eastern Europe in all its freshness and novelty. . . . In other words, Eastern Europe functions for the West as its Ego-Ideal: the point from which the West sees itself in a likeable, idealized form, as worthy of love. The real object of fascination for the West is thus the gaze . . . (Žižek 1992: 193).

My university's law school agreed to participate in the return gaze. Months later, our Soviet colleague arrived. This was in a rural northeastern college town, and our landscape was instantly transformed by her expressions of enchantment. Natasha (as I will call her) was from Moscow, and on her first evening she pronounced our campus with its old stone buildings and groves of trees as beautiful as a Baltic city of old. She was an energetic and thoughtful observer; we were flattered by her interest and engaged by her questions.

As she related it, Natasha's sense of mission had many components, and one of these involved making as broad a survey as possible of local institutions of law-making and law enforcement, among many other things. In aid of this project, my husband called an acquaintance who was a county court judge in a nearby city, in the hope of arranging a visit for Natasha on a day when the court would be in criminal session. It would be a day trip—a day out on democracy's road. When the day came, the three of us drove out together to listen to the morning's dramas. After half an hour or so, during a break, the judge sent the bailiff over to invite Natasha to tour the women's side of the county jail. She said she would be delighted, and she asked me to come along. We followed the bailiff into the lock-up area next to the courtroom just as a group of third graders filed out from their tour, looking a little dazed.

Going through that same door, we retraced the children's steps. The lock-up itself was a single large cell packed with men, standing or sitting silently on benches along the walls and on the floor; although it was winter, it was hot inside and the men were sweating. The cell was cinder block on three sides, bars along the corridor. Beyond the bars, the corridor wall was decorated with photographs torn from magazine centerfolds showing naked women in a variety of poses, cavorting at eye level. We looked at the men as we walked through, and they looked back. I had the impression that not just the pin-ups, but everything, everyone, could be sucked into the vortex to make these men feel their confinement to the utmost, and that we were now part of the prison. But the passage took only a moment. The bailiff walked us down that corridor and delivered us to a guard; soon, we were on our way again.

The guard explained jovially—I cannot recall the reason—that the only access to the women's jail that day was through the men's jail. This was not the ordinary route, he was saying as he escorted us up a flight of stairs into the men's jail. He was just ahead of us, turned perhaps 45 degrees back, toward us, smiling and speaking; we followed, smiling too, poised for polite conversation—as if what we were doing was walking somewhere together. But up in the men's section, it was even hotter than down below. At first, my main impression was of the heat, and this distracted me from whatever our guide was saying at that point. My glance wandered and fell on the eyes of one man. I could not look away right away; his eyes were simply miserable—utterly sad. He was not looking *to* us, as if inviting some response, but—from the darkness at the back of his cell—toward us. At that instant, I felt my freedom as a wall of embarrassment between us. If I looked uninvited, either I was an intruder or he was an animal in a cage; if I did not look, either I was in league with the warden or he was not a person. When I did look away again, I must have lowered my eyes, since it was only in that next instant that I saw that the man was completely naked. Then I saw them all: the men, completely or partially stripped, stood or sat in small groups, several men to a cell, utterly exposed to each other and now to us in whatever they were doing at the moment. None of them seemed startled, or made a move to cover themselves, but now Natasha and I looked away simultaneously, to the wall opposite the cells, a perfectly useless performance of unseeing on our parts.

Still, regaining my sight rearranged my other senses, and now I was aware of a smell that was unlike anything I had ever encountered. The smell did not come from the men, as it had downstairs, where the smell of sweat distinctly came from the men. But not here: It wasn't sweat, or human soil, or chemicals, or rotting food. It just was, and it filled the entire space. The men stood in it, in their stillness. We walked in it, although it made the distance to my own feet seem long and questionable. Our host chatted on amiably; I left the conversation to Natasha.

At the end of the row of cells there was an office, and we were invited in; we were to be handed over to a women's guard, and we would wait for her there. There was a pair of Norwegian criminologists there; they were just finishing up their research in the files, preparing to leave. There were introductions, we were offered chairs. The officers had just settled into a hospitable leisure when an alarm sounded. Then the telephone rang. One of the men in the cells had made a lasso out of a straightened coat hanger, with a razor blade wired to its end; now he was twirling it, threatening to cut anyone and everyone, and probably himself, too. There was no haste or sign of disturbance in the officer's response. Nothing unusual, he said. Purposefully, but without hurrying, he opened a desk drawer and pulled disposable rubber gloves from a box. "Because you never know," he explained; then he added as if confidentially, "AIDS," since our reactions had evidently been too slow for his sense of conversation. Cordially, he said that we would be perfectly safe where we were, since the doors would be locked from the outside and the glass was unbreakable. No cause for alarm. Click.

Now it was just Natasha and me, listening and not speaking, waiting. It began to seem like a long time. I remember feeling a powerful desire not to be there, and this surge of feeling must have shown in my expression. Natasha looked over at me and with a smile—a perfect guest reassuring her host—she said: "It's okay, Carol. The smell of unfreedom is the same everywhere."

This is how I learned it had a name. The smell of unfreedom clung to the officers as they came back to their desks—no one had been hurt; someone had talked him out of it—and the smell of unfreedom was in the corridor along which we finally made our way, and again on the stairs, as we approached the women's jail. The women's side was crowded, too, but it was cooler and airier, the cells brighter. All of the women were dressed, some of them were eating lunch from trays, some were watching television, some were talking. You could breathe. The guards, our guides, talked about drugs, prostitution, crowding. We saw the library, the classroom, the lounge, everything. We walked down more corridors. Time to go. Lunchtime. We were almost out the door, when the guard asked as an afterthought, "Do you want to see a solitary cell?" Natasha said yes. So we backtracked a little, walked down more corridors, and then another, and finally came to a heavy metal door with a small sliding panel at eye level. Natasha looked in, looked for a long time. Then she stepped away, silent, giving me a turn at the opening.

Conditioned by the conventions of the hotel industry, I expected the cell to be vacant. It was not. A young woman—African American like almost all the other inmates we had seen that day—was deeply asleep on a cot, tenderly cuddling a beautiful infant in the crook of her arm. Perhaps he had been nursing; now he slept with wet lips nestled against her cheek. He was three months old, the guard said. Mother and son had been given the solitary cell for its relative quiet and privacy. There was no irony in her voice as she said this, nor when she added that at the age of ten months the child would have to be placed in another home while his mother finished her sentence; that was the rule. Her sentence had a few years to go.

Natasha peered in again and asked the guard some questions. Then I looked again. There was no performance of unseeing here, as any audience for it was asleep. Having seized their privacy, I felt obliged to reciprocate with hope. I began an imaginary litany.

I imagined that they had a home outside the jail; imagined a father, aunts, uncles, grandparents, and siblings waiting for the baby at home to nurture him; imagined he would learn about being African American from them, and not from white policemen; that his body would be his own, not used against him as a humiliation; that he would always have and give love, never want for a warm bed or enough to eat. He had these things today, and perhaps he would forever be able to live the bond of tenderness with his mother during these months even though he would have no memory of her embrace. I imagined that he would have the pleasure of knowing his mother as he grew older, and she him; that he would have choices, and that his choices would lead him far, far away from this jail, from the cells on the other side of the wall.

Don't go there. I mentally separated that world from mine, and mentally claimed the boy for my world. This meant suspending what I knew about the fever pitch of public demands for law and order;[2] it meant setting to one side the fact that the United States has the highest incarceration rate in the world, that half of all criminal defendants found guilty are sentenced to jail; that the length of jail terms is growing; that crime is down but the number of acts designated as crimes is up, especially so-called victimless crimes like drug use, homelessness, and prostitution; the fact that depending on what you read, 25 to 50 percent of inner-city black men are in the criminal justice system (in jail, in prison, or on probation or parole), a figure that roughly matches unemployment in the poorest areas of the inner-city now that the jobs have moved off-shore or to the suburbs, and full-time work only deepens the poverty of men and, even more quickly, the destitution of women with children at home. The drug economy does not pay well, but it pays a little toward the American dream; the highest paid extra-legal waged work is prostitution.[3] But I don't know why this boy's mother was doing time. Given the length of her sentence, she might have been a federal prisoner housed in the local jail to ease the even greater crowding up the line. And be that as it may, she might have harmed no one, might even have been innocent. My litany, in retrospect, suspended her story altogether.

Absorbed in the question of this child's future, I told myself that there could—would—be real choices awaiting this boy when he was old enough to think and act for himself and the others he would care about. There would be an education waiting for him, too, and a job. I told myself he was a citizen, swaddled him in due process and the goodwill of his neighbors, employers, and fellow citizens, and saw him off into his future.

Much later, thinking it through, I realized these had not been hopes at all, but prayers. Thinking about that baby, my prayer had been that the odds would be with him; it was not hope simply to wish he might beat the odds. Neither option felt acceptable; both felt very unacceptable, in fact, since wishing and hoping in those terms only reproduce the statistical calculus of modern-day corrections for one's personal comfort.

Is there a student of social science who has not encountered Durkheim's famous formulation of the power of social judgment? Durkheim writes: "We must not say that an action shocks the common conscience because it is criminal, but rather that it is criminal because it shocks the common conscience" (Durkheim 1933:81). The context is *The Division of Labor in Society* (1893 [1933]), Durkheim's treatise on social solidarity.[4] Written on the eve of the Dreyfus affair, the book unfolds from his elucidation of two types of "conscience"—the individual and the collective. Both types of conscience always exist, Durkheim says, since individuals are always thinking for themselves even when social convention, the conventions of social description, or law says they are behaving according to the norm (Durkheim 1933:4–5; 1951:315). To put this another way, Durkheim sees the normal condition of society as multiple overlapping normative spheres—but he also recognizes some societies acknowledge this more fully than others; some might acknowledge it not at all. He refers to the question of how people think about their associations with others with the term *solidarity. Organic solidarity* is his term for associations that span normative communities; *mechanical solidar-*

ity, in contrast, refers to associations that acknowledge their common membership in only a single normative order. For Durkheim, then, the question of solidarity always involves a question of authority, at least insofar as the power to call someone to account for his or her actions entails recognition of multiple normative systems or only one—Durkheim's central thesis being that there is a direct relationship between organic solidarity and individual autonomy, given the way the recognition of multiple normative systems heightens an individual's self-awareness as a moral actor and buffers or mediates the internalization of any one normative discourse arising from the pronouncements of judges or other public authorities.

Durkheim's opening questions in *Division of Labor* are: "Why does the individual, while becoming more autonomous, depend more upon society? How can he be at once more individual and more solidary?" (Durkheim 1933:37; cf. Durkheim 1928). His answer is that individuals are paradoxically freer when they are more involved, since the differences among groups lead individuals to an awareness of their own choices.[5] By virtue of the institutional diversity inherent in modern life (he argues), modernity exhibits organic solidarity—but importantly, Durkheim offers this as a comment on the conditions of ethical responsibility, not as a benchmark of social evolution or modernity itself.

This said, *The Division of Labor in Society* (in his own times as in ours) has been received as a book about the evolution of diversity out of more homogeneous collective mentalities, as if Durkheim's distinction between mechanical and organic solidarity described some developmental threshold between "primitive" and "advanced" societies. The debate hinges on the question of what Durkheim might have meant by the phrase "collective conscience." Durkheim's contemporary critics tended to read his book as giving prominence to "collectivism," tending (as they do now) to hear "collective conscience" as *unanimity.*[6] Today, this reading has become conventionalized as a reception tradition that renders the book as a treatise on repressive law and the constraining effects of consensus[7]—in effect reading "common conscience" and "collective conscience" as consensus or "slavishness to custom" (Malinowski 1926: 5). This means that they pass too quickly over Durkheim's reflections on what might today be called the hegemonic aspects of discourse (for example, compare Giddens 1971:67) and his appeal to academic sociologists to challenge hegemony. Durkheim makes his own argument against these readings in terms that emphasize the analytical importance of understanding the conditions under which an individual is more or less likely to realize him- or herself as a moral agent (see note 6, above). For this reason, I read "collective conscience" as a reference to the aggregate moral universes of which any individual might be aware, and in relation to which his or her moral judgments take shape. What is repressive about mechanical solidarity is not the absence of difference but the lack of a discourse and institutional apparatus through which an individual might recognize his or her personhood outside of the dictates of status norms. Or, to put this differently, Durkheim locates the repressive element of mechanical solidarity in the way it precludes an individual's self-awareness. In contrast, the liberation of organic solidarity is in the way the public recognition of social and normative diversity fosters an individual's awareness of him- or herself amongst others, and accordingly, to effective moral reflection and deliberative exchange. Again, this is why I understand "collective conscience" in terms of the knowledge and other communicative demands of social action rather than some a priori claim about interdependence, consensus, and conformity.

In *Division of Labor*, Durkheim reflects on his own question by means of a sociological argument. He begins by inverting the conventional assumption (conventionalized under the banner of the social contract) that individuals must yield up their autonomy for the sake of the public interest, showing instead that "individual conscience" together with the conditions that make self-expression "audible" enhances the solidarity of society as a whole. In chapter 2 (the context of the lines quoted at the beginning of this section), he challenges the assumption that the public interest and universal notions of good and evil find full expression in penal law by a series of examples illustrating major sources of harm that go unaddressed in criminal law—the agents of financial ruin, for example, or lost reputation. Having shown that criminal law is selective in its attention to values, then, he goes on to consider the agents of that selectivity. His next step, then, is to demonstrate a connection between crime (i.e., acts that are designated as crime) and the legitimation needs of public authority.

Durkheim posits that the categories of *crime* do not derive from some universal scale of evils or public consensus as to interests, but from the way systems of authority make themselves known and maintain themselves "in some way transcendent" (Durkheim 1933:84). Specifically, he suggests that the category *crime* and penal law are the hallmarks of an authority claiming to act in the name of collective opinion—that very claim being the claim to "transcendence" referred to above. The interests of authority and its needs for self-legitimation determine *crime*, then, not the nature of the acts in question. Therefore (Durkheim continues), punishment is the palpable confirming sign of an order of authority, and a society that was free of crime would fall into chaos, since it would be bereft of the signs of its own existence *as* an authoritative order (Durkheim 1938: xxviii; see also Durkheim 1933: ch. 2, esp. pp. 84–85). Crucially, Durkheim treats social order as constituted in signs (by which society becomes thinkable for individuals) rather than as an institutional apparatus with particular effects. Criminal punishment features in the hermeneutics of authority, and it is for this reason that Durkheim can approach his main question of freedom as a problem of knowledge.

For Durkheim, in other words, crime and punishment are (to borrow from another context) "good to think" in the effort to understand—whether as sociologist or citizen—the nature of freedom. This is the context in which Durkheim gives the name *collective conscience* to the overall relationship between solidarity and self-recognition. Organic and mechanical solidarity represent different risks in this regard—in that these represent different degrees to which social orders are actively maintained on the basis of multiple or singular normative orders.[8] Mechanical solidarity, by virtue of claiming a monolithic normative order, implies the infinite fungibility of personal judgment. Organic solidarity, based on the interconnectedness of multiple normative orders, implies the necessity of self-expression.

But the modalities of solidarity are not, for Durkheim, ideologies or social charters. He maintains that collective or common conscience is accessible to social science because it is a "social fact"—"produced by collective elaboration" (Durkheim 1933:5). A "social fact" is "a thing" in that—like a material object—it can be known only by external observation, not "by even the most careful introspection" (Durkheim 1938:xliii). Then, when Durkheim defines "collective conscience" or "common conscience"—key terms throughout his corpus—as "the totality of beliefs and sentiments common to average citizens of the same society forms a determinate system which has its own life" (1933:79), he is characterizing a social fact in this sense. His reference to what is "common to average citizens of the same society" should therefore be read as specifying a social context, not as a claim to the uniformity of people's views (in that case "totality" would be redundant). Indeed, he stipulates that the collective conscience is contingent and diffuse, not a substrate; further, he specifies that law exists "outside" the common conscience. Immediately following the passage that opens this section, Durkheim continues:

> No doubt, [collective conscience] has not a specific organ as a substratum; it is, by definition, diffuse in every reach of society. Nevertheless, it has specific characteristics which make it a distinct reality. It is, in effect, independent of the particular conditions in which individuals are placed; they pass on and it remains. . . . As the terms, collective and social, are often considered synonymous, one is inclined to believe that the collective conscience is the total social conscience. . . . *Judicial, governmental, scientific, industrial, in short, all special functions are of a psychic nature, since they*

consist in systems of representations and actions. They, however, are surely out-
side the common conscience (Durkheim 1933:79–80; emphasis added).

Durkheim's elaborations of these ideas are in terms that are to some extent nowadays familiar—for example, his reference to "institutions" as social
facts; and to social facts as "collaborative" productions. His discussion of issues of knowledge and recognition develops further the linkage he draws between freedom and social knowledge: the essence of the social inheres in the
effects of believing that others know what one knows (mechanical solidarity)—or of knowing that they do not know what one knows (organic solidarity). This is what gives the collective conscience its coercive power (see
Durkheim 1938:ch.1)—since the collective conscience is sustained (in its
varied expressions) by the extent to which others' judgments are made out
to be substitutable for one's own. Each of Durkheim's prefaces to the various
editions of *Rules of Sociological Method* develops further this question of the
individual's relationship to the collective conscience, assigning to sociology
the task of examining the stakes in the highly unstable borders between inside and outside knowledge. *Because* this border is intrinsically unstable
(contingent, again, on the form of solidarity), sociology must draw a distinction between the social and the universal (Durkheim 1938:6–7).[9] Personal agency and autonomy are contingent social relations, not projections
of individual intention or choice; therefore, for Durkheim, it is *social science*
that articulates the collective conscience as a register of public discourse.

But in other crucial respects, Durkheim's language is unfamiliar. His
collective conscience might today be translated as *discourse*—the alignment of
words and worlds such that words have power, and individual subjectivities
are inseparable from the circuitries of knowledge that shape their milieus.[10]
Ideas—even highly personal ideas—have no clear point of origin, yet can
have powerful, even coercive, effects; indeed, from the standpoint of discourse, there can sometimes seem to be no clear distinction between the individual person and the collective social body—and this can be humanizing
or terrifying under different circumstances, as Durkheim also notes
(Durkheim 1951:315). Durkheim, like Foucault and others later, insists that
discourse is not a matter of language alone, but also of effects in which there
might be very high stakes for individuals or whole societies:

> Usually when collective tendencies or passions are spoken of, we tend to
> regard these expressions as mere metaphors and manners of speech with
> no real signification but a sort of average among a certain number of in-
> dividual states. They are not considered as things, forces *sui generis* which

dominate the consciousness of single individuals. None the less this is their nature . . . (Durkheim 1951:307).

Moreover, "collective tendencies . . . are forces as real as cosmic forces, and they affect the individual from without . . ." (Durkheim 1951:309). Still, the relationship between the person and the collective conscience is unsteady and highly personal, and collective conscience by definition cannot be determinative:

> Of course it is true that not all social consciousness achieves such externalization and materialization. Not all the aesthetic spirit of a nation is embodied in the works it inspires; not all of morality is formulated in clear precepts. The greater part is diffused. *There is a large collective life which is at liberty; all sorts of currents come, go, circulate everywhere, cross and mingle in a thousand different ways, and just because they are constantly mobile are never crystallized in an objective form* (Durkheim 1951:315; emphasis added).[11]

Modern readings tend to put all of this the other way around. Most readings make Durkheim's vision of law an extension of his concept of collective conscience, the latter conceived as consensus.[12] The main reception tradition makes "collective conscience" a system of *shared beliefs* (most recently, see Cotterrell 1999:53; see also works cited in note 4). Similarly, most readings make Durkheim's word "determinate" (in the definition of *collective conscience* quoted above) the equivalent of *determining* or *determinative*. To be sure, Durkheim posits a rhetorical connection between what he calls "the link of social solidarity" and "repressive law." The reception tradition has tended to make this connection into a straightforward matter of normative consensus, legitimation, and coercion. In an interesting departure, Cotterrell (1999:31) states that for Durkheim law is an "index" of solidarity, that is, the sign by which different forms of solidarity can be discerned and assessed in practice. In Cotterrell's words, law is "a mark of something more sociologically significant than itself" (1999:32). (Later, Cotterrell glosses Durkheim's view less in terms of a mark than mirror: "In early or simple societies, law and morality are both largely expressions of collective beliefs or sentiments rooted in religious sources" [Cotterrell 1999:53 ff.].) But it is, indeed, the indexicality of law (not its instrumentality) that Durkheim emphasizes in *The Division of Labor in Society.* He explains that the category *crime* signifies a distinction between approval and punishment—and moreover that it signifies a connection authorities *claim* (in the interests of their own legitimacy) between common conscience and law. Crime, therefore, can be usefully investigated as a way of exploring the self-legitimation of authority. Accordingly, Durkheim explores crime and punishment as a way of examining authority and solidarity—not the reverse (as would be possible if law simply "reflected" social attitudes and norms).

If one wants to understand the relationship between solidarity and the indexicality of law, the first task, Durkheim explains, is to understand "more precisely . . . what crime essentially consists of." In chapter 2 of *Division of Labor*, Durkheim begins that exploration by reviewing and rejecting in turn successive definitions of crime, a rhetorical tactic of ethnographic defamiliarization. Some paragraphs later, he concludes that crime originates not in the inherent criminality of particular acts or in the substantive public interest, but in the categorical opposition (at the level of a symbolic opposition) between approval and reproof—an opposition that is part and parcel of any system of authoritative order.

Later, in his discussion of mechanical and organic solidarity, he considers different formats for authority—as monolithic (mechanical solidarity) or plural (organic solidarity), and he associates different models of law with these (repressive and restitutive, respectively). But at this point in the chapter, he sets his task as challenging the assumption that approval and reproof refer to stable domains. Elaborating, he offers the lapidary sentence quoted at the outset of this chapter. He follows it immediately with a more straightforward version: "We do not reprove it because it is a crime, but it is a crime because we reprove it" (Durkheim 1933:81).

Durkheim's formulation remains a touchstone of modern social science, pointing simultaneously to the profound power of social judgment and its cultural contingency. The rhetorical power of Durkheim's passage is in its exposure of the subjective foundations of legal judgment, that is, the foundations of legal judgment in a symbolic order specific to a particular time and place—not universal morality. Moreover, he insists (in the same context; cf. pp. 77–82) that "collective conscience" is internally inconsistent (yet another reason to distinguish it from a collective attitude or mentality). Durkheim's "collective conscience" is not about uniformity, but rather the dilemma of how plural and (often) private discourses are acknowledged within any society's authoritative institutions. His formulation—far from making law an extension of public opinion—makes all such claims suspect. In France then, as in the United States now, criminal sanctions included the death penalty—a context in which the disparity between the irreversible finality of legal judgment and the potential frailty of subjective judgment delivers a certain shock. Durkheim's expository mode deploys such moments of shock and suspicion to widen the space for social science.[13] For example, in his discussion of the potential arbitrariness of social judgment, he emphasizes that such judgments are inconsistent, even contradictory. He refutes the commonplace notions that the scale of punishments reflects some scale of evils. Indeed, he dismisses these views with a counter-example, comparing the limited harm of a single homicide to the infinite harm of economic plunder: "In the penal law of most civilized people, murder is universally regarded as the greatest of crimes. However, an economic crisis, a stock-market crash, even a failure, can disorganize the social body more severely than an isolated homicide. No doubt murder is always an evil, but there is no proof that it is the greatest of evils" (1933:72). Nor are criminal acts necessarily the most damning, in his view: "We have not defined crime when we say that it consists in an offense to collective sentiments, for there are some among these which can be offended without there being a crime. . . . It is in like case with the reflections upon a woman's honor accruing from promiscuous intercourse outside of marriage. . . ." (1933:77). By examples such as these, drawing on conventional attitudes and defeating or qualifying them with examples drawn from everyday experience, Durkheim coaches and coaxes—disciplines—his readers to enter into the sociological method (see also Durkheim 1938: ix).

Durkheim's sociological method is especially vivid in *The Division of Labor in Society;* the readers—perhaps surrogates for the students who heard the original lectures—are actively addressed (in Durkheim's consistent use of "we" throughout the work, as well as in other ways). Importantly for my purposes in this essay, the dialogue he charts would take place in two registers: sociology and common sense.[14] The sociological element of exactitude is part and parcel of the rhetorical tactic of shock, as in the passages quoted above and others like them—the shock of evidence or contradiction against taken-for-granted assumption; it is only at such moments of disorientation that one can experience the *need* for social science in a personal way. This raises an interesting question of interpretation, since while Durkheim is clear that objectivity is the sociologist's *object* (rather than his or her *attribute*), he does not specify what he means by "objectivity." In the passage quoted, Durkheim states that meticulous description and scrupulous classification are necessary but not sufficient; the "objective element" in description and classification must be "discover[ed]"—*found* and *seen* to be the evidence with which to build a scientific argument (with "exact determination, and if possible, measurement"). Objectivity, in other words, is not a way of seeing, but the thing seen, from a vantage point in a mounting scientific argument.

What is the vantage point of that optic? Durkheim does not answer this question in so many words, but rather in the way he constitutes the dialogue between author and reader.[15] His articulation of the sociological method takes the form of direct appeals to the reader's capacity for fairness and re-flection. It would seem that if social science is objective it is not because ob-jectivity is some inherent *property* of the sociologist's activity, but because it defines *a relation* between the sociologist's discernment and alternative modes of knowledge. Scientific objectivity, in other words, is contingent on ethical dialogue and debate, on ethical judgments rooted in social conven-tion and personal experience yet open to revision in the light of knowledge of the vagaries of the human condition. Durkheim's argument makes objec-tivity and reflective practice inseparable.

Thus, when Durkheim indicates that description and classification are necessary but not sufficient elements of science, he implies a qualification: Meticulous description and scrupulous classification are *scientific activities* only to the extent that they operate within ongoing ethical dialogue. Objec-tivity does not define science; science is made by both its methods and its purposes—fundamentally pedagogical purposes. And similarly, it is not sci-ence that "makes" other people objective, but rather their desire for ethical dialogue and their openness to being informed by social science knowledge. Furthermore, the implication of Durkheim's passage is that science is strictly speaking retrospective, since it meets the criteria of science only at the in-stant it awakens fresh reflection.

Durkheim's scientific method, in other words, locates social science on an always-unfolding path within the gap between prejudgment and reflec-tion, attitude and experience. The scientific method is scientific within a tacit compact between the social scientist and the reader whose mutual con-cerns, candor, and good faith make it possible to stipulate that discovery is always also an ethical project. The social scientist's side of the conversation fulfills a particular responsibility stemming from both the nature of socio-logical authority *and* the conventions of everyday life. The reader's side of the conversation reflects a particular need for information at the very junc-tures where prejudice blocks constructive reflection yet remains open to re-vision. To put this differently, the sociological method is a scientific method to the extent that it marks the very edge of discourse—the very edge where fresh reasons might change the shape of reason itself.

Durkheim's construction of objectivity implies the existence of a non-discursive domain—the domain where "individual conscience" is distinct from the collective conscience (see Durkheim 1933:80 n. 9). This domain is by definition not unified—residing as it does in unique individuals; how-ever, it can be reclaimed for collective discourse through social science; in other words, the edge of discourse is the horizon between personal experi-ence and sociology. The "objective element" of sociology is at that horizon, ready for claiming by the sociological authority for the sake of convincing (never coercing) students and other readers.

As a case in point, what is the specific objectivity that Durkheim offers in the famous passage that opens this essay? From what distance does he hail us with those lines? I quote them again: "We must not say that an action shocks the common conscience because it is criminal, but rather that it is criminal because it shocks the common conscience." The phrasing has the makings of a constructionist charter—that is, an approach to crime that begins in the question of how discourse is organized around notions of the good and the bad, punishers and punished, and how these ideas are woven into the fabric of social recognition and legitimacy. He is asking us, to be sure, for an ethnography of crime, but he is also implying an ethical standpoint from which an ethnography of crime may be said to be objective and scientific. In the search for this location or starting-point, it is not just ethnography that is at stake, but the very possibility of social description, even society itself.

My own reading of Durkheim's passage was, until recently, more or less along the lines of the main reception traditions, with the addition that I saw the specific lines at issue as the beginnings of a deconstructive critique that might charter an ethnographic effort to understand how and where people's attitudes toward criminality are enmeshed with other elements of their social discourse, experience, and needs. However, when my long afterthoughts about what Natasha and I had seen in the jail gave me fresh reasons to return to this particular book, I felt (upon re-reading) that Durkheim's message was a more difficult one, something unremarked. Specifically, when Durkheim anchors the scientific method to the very limits of discourse, he recognizes that that boundary is a moveable one, precisely because it is constituted at the fundamentally fragmented and unexplainable horizon of individuals' unique responses to the world. Crucially, though, this is not an individualistic claim, but a statement about the inherent ambiguity of the border between the inside and the outside—and correspondingly for anyone, the problem of situating one's own agency (or realizing it at all) in relation to the contingencies of one's own social knowledge.[16] This implication is clearer in *Suicide:*

> These tendencies of the whole social body, by affecting individuals, cause them to commit suicide. The private experiences usually thought to be the proximate causes of suicide have only the influence borrowed from the victim's moral predisposition, itself an echo of the moral state of society. To explain his detachment from life the individual accuses his most immediately surrounding circumstances; life is sad to him because he is sad. Of

course his sadness comes to him from without in one sense, however not
from one or another incident of his career but rather from the group to
which he belongs. This is why there is nothing which cannot serve as an
occasion for suicide (Durkheim 1951:300).

Change the suicide's "sadness" to society's judgment of the criminal—and
there, I believe, is the challenging essence of Durkheim's passage in the earlier book.

If crime, as he says, is the name for the broken tie between social solidarity and its symbolic expression in law, then sociology's task is not to restore or repair the tie by seeking the positive link between society's
judgments and the men and women who are disqualified from society by—
and even before—their convictions. Rather, the task is to invigorate social
solidarity anew, with a fresh social compact built of fresh facts.

In referring to the smell of unfreedom as universal, I understood Natasha's
comment as cosmopolitan. I took her to mean that this jail was behind some
false paving stone; lifting it opened a gaping global underworld to view.[17] Indeed, it was something like a sense of falling into a place without geography—where East and West, democracy and the Soviet system, were not
distinct systems or locations, but akin to reports of changeable and distant
weather, equally irrelevant, as if overheard in a place without windows.[18]

When we were back outside, with the judge and my husband, Natasha
pronounced the jail "not bad," except for the absence of fresh air and natural
light, and any useful occupation for the inmates. She was critical of the pin-
ups; the judge thought she was thinking of the school tours. We told the men
about the woman and her baby in solitary; the judge took it as a comment
on overcrowding, and told us about the county's plans to build a new jail.

We left after that, and rather than turn homeward, the three of us decided to drive on, to Niagara Falls. We rode largely in silence. We were still
in the thrall of unfreedom when we arrived at the falls, but they are so mag-
nificent that we bestirred ourselves and made room for them. Natasha had
never been there before, but her grandfather had, in the 1920s, on a commission of some kind. She chose a vantage point for us from a photograph
that he had taken at the time, and that had been passed along as a small heir-
loom. She recognized the exact place immediately. We went to the spot,
watched the water, took turns with the camera. We ate hotdogs at a nearly
deserted café, and then we went back to the falls. The cold mist was surging,
clinging to every surface. As darkness fell, we left.

On the long ride home, the three of us began to talk. We did not talk about what we had seen; or rather, we talked about it by talking about other things. We traded stories, family stories, stories of our parents and grand-parents, their journeys, their characters, the stories of their names, their un-expected kinship (Fyodor and Dorothy both mean "gift of God"), and many miles later, when we were all too tired for stories, we traded sayings, and then words—just words, the most beautiful words we could think of, she in Russian and we in English.

In retrospect, this, finally, was hope. It was hope not because naming beauty could restore anyone to freedom or change that child's future, but be-cause it means you know you are part of a world that is shared. And in such a moment of choice or recognition (or something in between) the mind's eye sometimes yields an "objective element" that the light of day easily obscures: for each of us, the need for social science may be different, but from that in-stant of grasping the objective element, we can be drawn through it into a common conversation. Natasha and I had seen men, women, and a child in those cells that day, people whose names we did not know, whose crimes we had no knowledge of, who may have committed no crimes, whose pasts and futures will be forever blank for us, to whom we did not speak—and whom we had managed to acknowledge personally only by looking away. I would not wish to be misunderstood: This is not an ethnographic method. It is a confes-sion, for the sake of explaining why, in retrospect, stripped of the syntax of so-cial judgment, the presence of every one of the individuals in those cells was as incongruous as that baby's. That day, we had talked about the baby's odds, but there is also the pressing matter of the adults' odds, including the odds against the "objective element" for science and policy ever being found in their individual uniqueness, their wholeness, or even—if we are to limit discussion to what we could claim actually to know about them as people—their sadness.

Perhaps this sounds absurdly sentimental, or just beside the point. But it is how I came to consider that Durkheim's famous formulation of crime and punishment does not stop at the arbitrariness of social judgment. He takes science—and the ethical shock that makes it science—all the way to "the link of social solidarity" with the judged. The ethnography of crime does not begin with forbidden acts, arrests, and prosecutions. It does not begin inside the category of crime, in other words, but rather marks out and unsettles a domain beyond it, a domain where the starting point for social science can begin in the naked fact of fellow humanness. From that territory, uncharted except by anyone's readiness to enter it, the sociological method can lend it-self to the task of bending the alignment of conventional social description and the prejudicial discourse of social judgment. To say this is not to claim innocence for the social scientists, or to place them—us—in some impossible place apart in the realm of the good, but the opposite; inevitably, *contra* the very discourse in question, there is no place to live outside of the world of names and judgments. Perhaps this is why Durkheim does not ask us to fit the punishment to the crime, but to think again about solidarity as the epistemological and ethical basis for sociology's method and authority alike.

This brings us once again to the passage that is my main text in this chapter. Reading the passage as a comment on the force of consensus and the social basis of repressive law is misleading, in that it confuses "collective conscience" for homogeneity, and "determinate" for determining—in other words, confusing the indexicality of law in the abstract for the instrumentality of law in action. And reading the passage in constructionist terms is inadequate, too, since this would take us only so far as the question of how acts are labeled and judged, begging the questions of which people are most liable to be charged with committing them, how *they* are labeled and judged, and what our responsibilities as social scientists might be to them—whether or not we are *studying* them. I have offered the story of my day out with Natasha to explain why I now read Durkheim's lines as a statement about the centrality of rethinking crime and punishment to the project of recalibrating ethical dialogue about the value of social science itself.[19] I believe Durkheim marks the place for that dialogue at the border between freedom and unfreedom—since the horizon borne of the radical juxtaposition of conditions of freedom and unfreedom is the objective element of social science itself. It seems to me, at least, that the "we" of Durkheim's passage is no mere rhetorical flourish, but a demandingly inclusive first person plural that encircles social scientists, legal professionals, and the general public—and also each of the people inside the prison walls.

In pursuing this possibility, the ethnography of crime extends keenly homeward the critical insights of scholars whose ethnography in fields marked by violence, terror, and ethnocide has enlarged the conceptual power of ethnography and, accordingly, widened the range of ethnography as a genre of writing.[20] Their reassessments begin in a refusal to fix a boundary between inside and outside, and to refuse the social distance that is often (deservedly) an object of professional caricature—so as to keep closer track of the differentials and distances that pass for decorum or destiny in the world we all share. Or as Durkheim writes, contemplating the double-sidedness of collective conscience as both dangerous and nourishing: "A group is not only a moral authority which dominates the life of its members; it is also a source of life sui generis. From it comes a warmth which animates its members, making them intensely human, destroying their egotisms" (Durkheim 1933:26).

ACKNOWLEDGMENTS

I am grateful to my hosts and colleagues for the contexts that occasioned earlier versions of this essay—at the Law and Society Association, the Women's Caucus of Democrats Abroad (Paris), and the Department of Anthropology Colloquium, University of California at Irvine. For their encouragement to experiment with ethnographic genres, and their critical responses, I am grateful—though not for this alone—to Fred Aman, Sheila Malovany-Chevallier, and especially Stephanie Kane and Phil Parnell. I hope it is not too paradoxical an acknowledgment to express my debt to Natasha under the pseudonym I have given her in the essay.

NOTES

1. Cover 1986; also Sarat and Kearns 1992.
2. This was the year after Republican George Bush's presidential campaign deployed an advertisement against Democrat Michael Dukakis (governor of Massachusetts), featuring Willie Horton, a black man arrested for rape while on early release from a Massachusetts prison where he had served time for a previous conviction.
3. On the emergence of contemporary trends in criminology and penology, see Feeley and Simon 1992; Garland 1985; and Simon 1993, 1998. On poverty and the illegal economy, see Handler 1995. On the retail drug trade, see Bourgois 1994. On racism, see Butler 1993 and Omi and Winant 1994.
4. For assessments of *The Division of Labor in Society* in Durkheim's oeuvre, see Mauss 1928: esp. vi; Giddens 1971: Part II, 1978; Lukes 1973: Part II, esp. ch. 7; Filloux 1977; Jones 1986: ch. 2; Gane 1993; Turner 1993; Miller 1996: esp. ch. 3; Cotterrell 1999. For discussion of the relevance of the Dreyfusard movement to Durkheim's work on *Division of Labor, Rules of the Sociological Method,* and *Suicide,* see Lukes 1973: 332–354; Jones 1986: 19; and Miller 1996: 107.
5. Individuals' private responses and intentions lie beyond social science, however (Durkheim 1938: 95).
6. Durkheim answers his contemporary critics directly in his prefaces and introduction to the first and second editions of *The Rules of Sociological Method* (1938[1895]: esp. pp. xli, xlvii, lx, 9), as well as throughout *Suicide* (1951[1897]), which in important respects completes the demonstration initiated in *Division of Labor* (a pairing noted by Mauss 1928: vi; see also intellectual biographies by Giddens and Lukes, cited above).
7. See, for example, Giddens 1971: ch. 5, esp. pp. 74–75; Giddens 1978: 24; and Lukes 1973: ch. 7; cf. Filloux 1977: 3, as well as Cotterrell 1999, already noted.

8. Durkheim—committed to the Dreyfusard movement contemporaneously with his work on *Division of Labor, The Rules of Sociological Method,* and *Suicide* (Lukes 1973: 332–354; Jones 1986: 19; Miller 1996: 107)—seems to associate mechanical solidarity with a state that claims the monopoly on legitimate political expression. Such conditions can only lead to the progressive depoliticization of social life, he argues—that is, to losses in individual autonomy (i.e., moral self-recognition) as the discourses of norms and values are increasingly absorbed into the state sphere (Durkheim 1933: 5; see also Filloux 1977: 351; Miller 1996: 8–9).

9. On these same grounds, Durkheim is explicitly critical of positivism and functionalism, and he states that one purpose of *Division of Labor* is to provide a demonstration of this critique (Durkheim 1938: ch. 5; cf. Giddens 1978: ch. 1).

10. For further critical discussion of the reception tradition surrounding Durkheim's work, especially in the United States, see Jones 2001: esp. 5–21. See also Gane 1993.

11. On discourse, see Foucault 1973, 1978. Key elements of Durkheim's concept of collective conscience are developed in *Suicide.*

12. Cotterrell (1999), Giddens (1971, 1978) and others (see note 1, supra) suggest further that Durkheim regarded law as the extension of the common conscience; however, this would seem to be specifically excluded by Durkheim himself, as already noted.

13. The implications for social science methods are developed further in *The Rules of Sociological Method* (1895) and *Suicide* (1897).

14. On sociological method and common sense, see Durkheim 1938: xxxvii.

15. Indeed, his books are based on his lectures to students at Bordeaux and the Sorbonne. Education—and his own pedagogical practice—were also among his major sociological concerns (on Durkheim's pedagogy, see Lukes 1973).

16. Durkheim sometimes uses the phrase "*sui generis*" as shorthand for the fundamentally unexplainable force of an idea or relation. Cf. Appadurai 1996. On the relationship between globalization and poverty, see Sassen 1988.

17. I am grateful to Teresa Caldeira for pointing out that the world that made those distinctions seem like cardinal points of orientation has itself now changed.

18. Mary Poovey's recent essay examines the political economy of values currently shaping academic debates over the value of the humanities—concluding in part: "I think that the only way we can inaugurate a discussion about alternative definitions of value is to risk asserting that there are other goods that must exist if we are to remain human" (2001: 14). I am grateful to Bill Maurer for bringing Poovey's article to my attention.

19. For current and comprehensive discussion and resources in this rich and now extensive literature, see Das 2000 and 2001. On the issue of ethnographic genres and their responsiveness to contemporary insecurities, see also Benson 1993, Bhabha 1998; Clifford and Marcus 1986; Daniel 1996; Malkki 1995, 1997; Nordstrom and Martin 1992; Lavie, Narayan, and Rosaldo 1993; and Tsing 1993.

BIBLIOGRAPHY

Appadurai, Arjun
1996 Modernity at Large: Cultural Dimensions of Globalization. Minneapolis: University of Minnesota Press.
Benson, Paul, ed.
1993 Anthropology and Literature. Urbana: University of Illinois Press.
Bhabha, Homi
1998 Anxiety in the Midst of Difference. Political and Legal Anthropology Review 21(1): 123–137.
Bourgois, Philippe
1994 In Search of Respect. Cambridge, England: Cambridge University Press.
Butler, Judith
1993 Endangered/Endangering: Schematic Racism and White Paranoia. *In* Reading Rodney King/Reading Urban Uprising. Robert Gooding-Williams, ed. Pp. 15–22. New York: Routledge.
Clifford, James and George Marcus, eds.
1986 Writing Culture. Berkeley: University of California Press.
Cotterrell, Roger
1999 Emile Durkheim: Profiles in Legal Theory. Stanford, CA: Stanford University Press.
Cover, Robert M.
1986 Violence and the Word. Yale Law Journal 95: 1601–1929.
Daniel, Valentine
1996 Charred Lullabies: Chapters in an Anthropography of Violence. Princeton: Princeton University Press.
Das, Veena, ed.
2000 Violence and Subjectivity. Berkeley: University of California Press.
2001 Remaking a World: Violence, Social Suffering, and Recovery. Berkeley: University of California Press.
Durkheim, Emile
1928 Le Socialisme. Marcel Mauss, ed. Paris: Librairie Félix Alcan.
1933 (1893) The Division of Labor in Society. George Simpson, trans. New York: Free Press and London: Macmillan.
1938 (1895) The Rules of Sociological Method. 8th ed. Sarah A. Solovay and John H. Mueller, trans. George E. G. Catlin, ed. New York: The Free Press.

1951 (1897)   Suicide: A Study in Sociology. John A. Spaulding and George Simpson, trans. George Simpson, ed. New York: The Free Press.

Feeley, Malcolm and Jonathan Simon

1992   The New Penology. Criminology 30(4): 449–474.

Filloux, Jean-Claude

1977   Durkheim et le socialisme. Geneva: Librairie Droz.

Foucault, Michel

1973   The Birth of the Clinic. A. M. Sheridan Smith, trans. New York: Pantheon.

1978   The History of Sexuality, vol. 1. Robert Hurley, trans. New York: Pantheon.

Gane, Mike, ed.

1993   The Radical Sociology of Durkheim and Mauss. London and New York: Routledge.

Garland, David

1985   Punishment and Welfare: A History of Penal Strategies. Aldershot, Hants, England: Gower.

Giddens, Anthony

1971   Capitalism and Modern Social Theory: An Analysis of the Writings of Marx, Durkheim and Max Weber. Cambridge: Cambridge University Press.

1978   Durkheim. London: Fontana Press.

Handler, Joel F.

1995   The Poverty of Welfare Reform. New Haven, CT: Yale University Press.

Jones, Robert Alun

1986   Emile Durkheim: An Introduction to Four Major Works. Beverly Hills, CA: Sage.

Jones, Susan Stedman

2001   Durkheim Reconsidered. Cambridge: Polity.

Lavie, Smadar, Kirin Narayan, and Renato Rosaldo, eds.

1993   Creativity in Anthropology. Ithaca: Cornell University Press.

Lukes, Steven

1973   Emile Durkheim: His Life and Work: A Historical and Critical Study. Harmondsworth: Penguin Books.

Malinowski, Bronislaw

1926   Crime and Custom in Savage Society. New York: Harcourt, Brace.

Malkki, Liisa

1995   Purity and Exile: Violence, Memory, and National Cosmology among Hutu Refugees in Tanzania. Chicago: University of Chicago Press.

1997   Newsstand Culture: Transitory Phenomena and the Fieldwork Tradition. In Anthropological Locations: Boundaries and Grounds of a Field Science. Akhil Gupta and James Ferguson, eds. Pp. 86–101. Berkeley: University of California Press.

Mauss, Marcel

1928   Introduction. In Emile Durkheim, Le Socialisme. Paris: Librarie Félix Alcan.

Miller, W. Watts

1996 Durkheim, Morals, and Modernity. London: UCL Press.

Nordstrom, Carolyn and Joann Martin, eds.
1992 The Paths to Domination, Resistance, and Terror. Berkeley: University of California Press.

Omi, Michael and Howard Winant
1994 Racial Formations in the United States: From the 1960s to the 1990s, 2nd ed. New York: Routledge.

Poovey, Mary
2001 The Twenty-First-Century University and the Market: What Price Economic Viability? Differences: A Journal of Feminist Cultural Studies 12(1): 1–16.

Sassen, Saskia
1988 The Mobility of Labor and Capital. London: Cambridge University Press.

Simon, Jonathan
1993 Poor Discipline. Chicago: University of Chicago Press.
1998 Refugees in a Carceral Age: The Rebirth of Immigration Prisons in the United States 1976–1992. Public Culture 10(3):577–606.

Tsing, Anna Lowenhaupt
1993 In the Realm of the Diamond Queen. Princeton: Princeton University Press.

Turner, Stephen, ed.
1993 Emile Durkheim: Sociologist and Moralist. London and New York: Routledge.

Žižek, Slavoj
1992 Eastern Europe's Republics of Gilead. In Dimensions of Radical Democracy: Pluralism, Citizenship, Community. Chantal Mouffe, ed. Pp. 193–207. London: Verso.

EPILOGUE

Stephanie C. Kane

ICONIC PERPETRATORS HANG OUT ON THE COLLECTIVE CORNERS of belief, drinking muscatel. Crack addicts, mad bombers, and deadbeat dads: With some analytic distance, their faces and deeds appear replaceable, typological, interchangeable; they morph with time, culture, geography, technology. They're invented, contested, mass produced. Some are ignored. The ones who do the most damage are most likely to be protected by wealth and position, others go unnoticed because their victims are undervalued by racist and gendered regimes. Selected icons zip almost instantaneously, fiber optically, across ocean beds, under the lobsters and the crabs. A great global bricolage of deceit, revenge, and pathology gathers and focuses all the anxious rays, all the mundane indignities and senseless losses that smudge the selves, latching individuals to a directed, righteous anger: This is crime's ideological power.

The functional dynamic of crime's symbolic elements and patterns of narrative coherence are intricately embedded in our diverse world cultures. Its roots are as ancient as myth, its tendencies postmodern. Historically situated, its representation has the power to wrench our guts. It becomes critical for a person to intervene in this peculiar type of subjection, and to reject the common rationales that verify the status quo through manipulation, quantitative and qualitative significance notwithstanding. The discourse must be moved in new directions, if only in small rooms among small groups.

Ideological formations of crime are packaged, stamped with corporate logos, and sent forth into the planetary message stream like advertising. Real events interact with fictional scapes to generate intertexts and subtexts with casts of grisly characters. Before today's sensory overload threw empathy into question, the capacity to respond to the images of pain and fear that criminals cause may once have seemed almost universal, a kind of urban herding and hunting instinct, expressed in primitive ritual languages and whispered in the hallways of our legislatures and institutions of justice. But perpetrating icons are more accurately read as symptoms, not causes, of the profound disregard for suffering that humans have for all sentient beings, including themselves. This knowledge and its televised evidence—the famines, the epidemics, the weapons of destruction for minors and masses—these are too great, too hard to deal with. It is these from which we would like to be distracted. The iconic perpetrators are vehicles of this distraction, although at times they do accomplish enough damage to truly index large disasters all on their own—Milosevic, bin Laden, the Enron execs, to name a few in recent news. History, as an accretion of singular events, and public culture, as imaginative projection, (e)merge in these larger-than-life figures.

The perpetrator icons and the dramas they enact are mass media staples; the more sensational, the richer the spoils. Mass culture is thus tilted toward synchronized axes of extremes. The global addiction to crime engenders a reactionary and delusional political discourse on justice. It has, to take one example, made the United States, a nation saturated with all possible modes of mass communication, the *dueño* (owner) of the largest prison population in the world. Cultural hegemony is so tightly knitted to local interests yet so antithetical to them, that democratic governance must be fortified by militarization. The industrialization of forced confinement has become a vast and lucrative national endeavor that has led to the replacement of educational institutions by prisons, jails, and detention centers. The shift of resources reduces the number of informed participants in the democratic process and increases the number of convicted felons who've lost the right to vote. In this way, fascism can replace democracy without anyone having to stage a coup.

I'm not saying that real crimes don't happen to good people. But what happens next, or next to, or before? Why the exaggeration of some forms of crime and the misrecognition of others? What are the processes that situate crime, that incite it only to parasitize the fervor it provokes more efficiently? From whence the sadistic pleasures of punishment? However it enters the discursive currents and backwaters of our social fields, crime pricks our collective consciences and goads us to social action. It is a chain of signification that ties the unconscious to the conscience, the doomed trickster in the old-new world order. As such, it becomes focal points for argument and interpretation, signposts in relation to which people remember their life histories.

Crime is the focus *and* the frame; the shape-shifters, the stage and the props. And the frame is just as shifty as the shapes. The ethnographer's gaze searches beyond an event's immediate dimensions, stretching across knowledge horizons, moving beyond crime as an already-given category of act, person, or quality. Linking the empirical coordinates of material conditions and beliefs, this is a discipline for inter-relating myriad pieces of multidimensional, holistic puzzles, and to do so ethically, without building up divisive perimeters and objectifying exotic cultures. Ethnography is a good toolkit: method enough to establish a systematic and relevant database; theory enough to bring insight to bear; and room enough to allow serendipity to infuse the resulting text with the zeitgeist. Data collection is omnivorous, directed and open, progressive and recursive.

Keyed to world events at the level of social interaction, ethnography can be a kind of quest, for the method does allow one to put body and soul to test. One throws oneself into it—a place, a people, a crime—in a reflexive interpretive process that draws new dimensions of objectivity out from the subjective; that experiments with observations of convention, contest, and consensus, digging for what is not said, participating to get the feel of a certain kind of existence; waiting for unforeseen events that magnetize people's attention and reveal social structures, at the same time allowing the ethnographer and her notebook to inhabit the guise of a fly on the wall. There can be some sympathetic magic involved, some secret stirring that expands, creating a vibrational similarity of affect that animates the public sphere and generates novel alignments and crystallizations of interests. Yet this implies a concordance that, if it occurs at all, is evanescent and unpredictable. And crime is so paradoxical, justice so fickle. The social order can be so disordered, the line between right and wrong so obscure. Is it police brutality or a fluke accident of the swats just doing their job? Is it anthrax or the flu? Rape or consensual sex? The right categorization is often crucial yet difficult to discern. There are disagreements, cross-currents, falls through the cracks.

Ethnographic research design is aimed at complexity of understanding. Through the evocation of paradigmatic scenes of social life, ethnographers can reveal crime's power in all its vivid particularities and slippery personifications. Ethnographers enter the social field, document the conditions, and decipher the logics of exchange; the rules, hierarchies, and plots; the poetic, the common, and the eccentric. For the most part, the meaning of criminal events emerge from prior study and in the aftershocks. Rarely are we witnesses to the scene. But we can analyze the modes of crime: the conditions of its production and imagination, the dynamics of the larger struggle of which it is a part. While not as accessible for study as modes of production, the ethnographic study of modes of crime can lead to the development of useful frameworks for thinking about a justice system that can deal justly with our differences.

Unlike most other forms of scholarly inquiry in the social sciences and humanities, ethnographic fieldwork requires "being there," standing within one's subject, learning through sensation as well as mindfulness. The combination creates a unique vantage point from which to write about social reality. The authors of the essays in this volume identify a number of dimensions of crime's power that I daresay would not have been arrived at in quite the same sense by other means. They illustrate a number of overlapping themes that can be studied and developed in a broad range of cultural settings. For example:

- As symbols, crime and criminality are basic structural elements in the dynamic interaction of myth and history (Kahn).
- Criminal organizations mediate relationships of inequality and oppression among corporations, governments, gangs, and communities. Their criminality may be exploitative, unstable, reversible, and/or service-oriented (Nader, Wedel, Parnell).
- Criminality manifests itself in different guises. The expression and suppression of the guises of criminality, like revolution and war, are often controlled by the patriarchal powers and upper classes of nation states (Martin, Starr).
- The state's violent criminal coercions rely on ritualized performances of humiliation. Its power is fluid, moving from dope dealers to diplomats, from dictators to Chambers of Commerce (McMurray, Linger).
- The paintings or the writings of serial killer-types can create high/low culture flaps. Crime is a commodity (Brydon and Greenhill).
- In their very attempts to analyze crime, ethnographers become implicated in its power. This provocation leads to a rethinking of the ethnographer's role in history and society (Mark, Greenhouse).

As an open set of possibilities with which to further the ethnography of crime as a scholarly endeavor, a collaborative "confrontation at the meeting place of difference," to use Phil Parnell's turn of phrase, we offer an array of principles and processes, and the scenes of social life from which they are drawn.

CONTRIBUTORS

ANNE BRYDON is Associate Professor of Anthropology at Wilfrid Laurier University in Waterloo, Canada. Her current writing engages with modernity and modernism in Iceland. With Sandra Niessen she co-edited *Consuming Fashion: Adorning the Traditional Body.*

PAULINE GREENHILL is Professor of Women's Studies at the University of Winnipeg. Her most recent book, co-edited with Diane Tye, is *Undisciplined Women: Tradition and Culture in Canada.*

CAROL GREENHOUSE is Professor of Anthropology at Princeton University. Her research and teaching focus on the ethnography of law in the United States. Her recent publications include *A Moment's Notice: Time Politics Across Cultures* and, as editor, *Democracy and Ethnography* and (with Elizabeth Mertz and Kay Warren) *Ethnography in Unstable Places.*

HILARY ELISE KAHN is on the faculty of the anthropology department of Indiana University in Indianapolis. Her ethnographic and video research with Q'eqchi' Mayan migrants in Livingston, Guatemala, is the basis of a collaborative video entitled "And with this We Live" and an article on identity politics published in the *Journal of Latin American Anthropology.*

STEPHANIE KANE is Associate Professor of Criminal Justice with appointments in Anthropology, Folklore, and Gender Studies at Indiana University, Bloomington. She is the author of *The Phantom Gringo Boat: Shamanic Discourse and Development in Panama* and *AIDS Alibis: Sex, Drugs, and Crime in the Americas.* She is currently writing an ethnographic biography of an ex-con.

DANIEL T. LINGER is Professor of Anthropology at the University of California, Santa Cruz. His writing focuses on culture theory, identity, transnational experience, politics, violence, cities, and face-to-face interaction. He is the author of *Dangerous Encounters: Meanings of Violence in a Brazilian City* and *No One Home: Brazilian Selves Remade in Japan.*

VERA MARK is an anthropologist and Assistant Professor of French and Linguistics at Penn State University. Her research and teaching center upon the study of everyday life and questions of identity as seen through the interfaces between regional, national, and global spaces. Her most recent article, co-authored with Joan E. Gross and published in *French Cultural Studies,* concerns the recasting of southern regional identity in France through the genre of Occitan rap music.

JOANN MARTIN is Associate Professor of Anthropology at Earlham College. Her ethnographic research in Mexico focuses on social movements, the politics of gender/sexuality, and memories of history. She is co-editor with Carolyn Nordstrom of the book *The Paths to Domination, Resistance and Terror* and is completing a book entitled *The Passions of Place: Global Politics in a Transnational World.*

DAVID MCMURRAY is Assistant Professor of Anthropology at Oregon State University. His ethnographic research on the Berbers in Morocco and Europe focuses on the politics of minority cultures, as well as the relation between commodity circuits, border runners, and the state. He is the author of *In and Out of Morocco: Migration and Smuggling in a Frontier Boomtown.*

LAURA NADER is Professor of Anthropology at the University of California at Berkeley. She is the author of several pioneering works in the anthropology of law, including *Harmony Ideology: Justice and Control in a Zapotec Mountain Village* as well as the edited volumes *The Ethnography of Law, Law in Culture and Society,* and *The Disputing Process: Law in Ten Societies,* co-authored with Harry F. Todd.

PHIL PARNELL is Associate Professor of Criminal Justice with appointments in Anthropology, Caribbean and Latin American Studies, and the Center on Southeast Asia at Indiana University in Bloomington. He is the author of *Escalating Disputes: Social Participation and Change in the Oaxacan Highlands* and is currently writing an ethnography about how the state, law, and crime are shaped by the lives of Manila's urban poor.

JUNE STARR held positions in anthropology and law throughout the world, including many years as Professor of Cultural Anthropology at the State University of New York at Stony Brook and more recently as Professor of Law at Indiana University Law School at Indianapolis. She is the author of *Dispute and Settlement in Rural Turkey* and *Law as Metaphor: From Islamic*

Courts to the Palace of Justice. She co-edited *History and Power in the Study of Law: New Directions in Legal Anthropology* with Jane Collier and *Practicing Ethnography in Law: New Dialogues, Enduring Practices* with Mark Goodale.

JANINE R. WEDEL is Associate Professor of the School of Public Policy at George Mason University and a fellow at the National Institute of Justice. She has studied eastern Europe's evolving social and economic order for more than 20 years. She is the author of *Collision and Collusion: The Strange Case of Western Aid in Eastern Europe, 1989–1998; The Unplanned Society: Poland During and After Communism;* and *The Private Poland: An Anthropologist's Look at Everyday Life.*

INDEX

Cunha, Euclides da, 114–15
Currie, Elliott, 56–7

DaMatta, Roberto, 114
Death Squads, *see* São Luís, terror-
squad abductions
Delegacia Especial da Mulher (DEM),
102–3, 118–19
democracy
in Brazil, 99
in Eastern Europe, 270
and Philippine grassroots land
organizations, 198–200, 214
and intergroup networks, 15–16
and Philippine land syndicates, 219
and Philippine urban poor, 200
among the Zapotec, 64–5
deportations of Jews from France, 253,
261–2
dictatorship, *see* repression, military;
Marcos;
Zapata, Emiliano; Zapatistas
dirty togetherness, 225, 226, 228
discourse
Durkheim on, 278, 280, 282–3
moral, 10–11
political, 11
public, 145–6, 278
see also media discourse
discourse analysis, 146–7
The Division of Labor in Society, 274–6,
279–80, 281
Durkheim, Emile
and authority, 19, 275–7, 279–80,
283
and coercion, 278, 279
and collective consciousness, 19–20,
274–80, 282, 286
and crime, 274, 276–7, 279–81,
283–4, 286
and discourse, 278, 280, 282–3
The Division of Labor in Society,
274–6, 279–80, 281

and ethnography of crime, 269–70,
283, 285, 286
and freedom, 20, 277–9, 286
and individual agency, 19, 275–6,
278, 283–4
and individual consciousness, 274,
278, 282
and law, 276–8, 279–81, 284, 285,
286
and legitimation, 19, 279–80
and mechanical solidarity, 274–7,
278, 280
and moral agency, 19–20, 275, 280,
286
and multiple normative systems,
19–20, 275
and objectivity, 19–20, 281–3
and organic solidarity, 19, 274–7,
278, 280
and the rhetorical tactic of shock,
280–1, 283, 285
and scientific method, 269–70,
281–3
and social facts, 277–8
and social judgment, 20, 274,
280–1, 285, 286
and social knowledge, 277–8, 283
and social solidarity, 274–6, 278,
279–80, 284
and sociological authority, 269–70,
282
and sociological methods, 269–70,
281–2, 286

Eastern Europe
and communism, 14, 222, 224,
226, 227–8, 238
and corruption, 225, 232, 234, 240
and criminal category, 14–15,
221–3, 232–3, 238, 240
and dirty togetherness, 225, 226, 228
and finagling, 224, 225
and flex organizations, 14, 230